# SAIC
## THE FIRST THIRTY YEARS

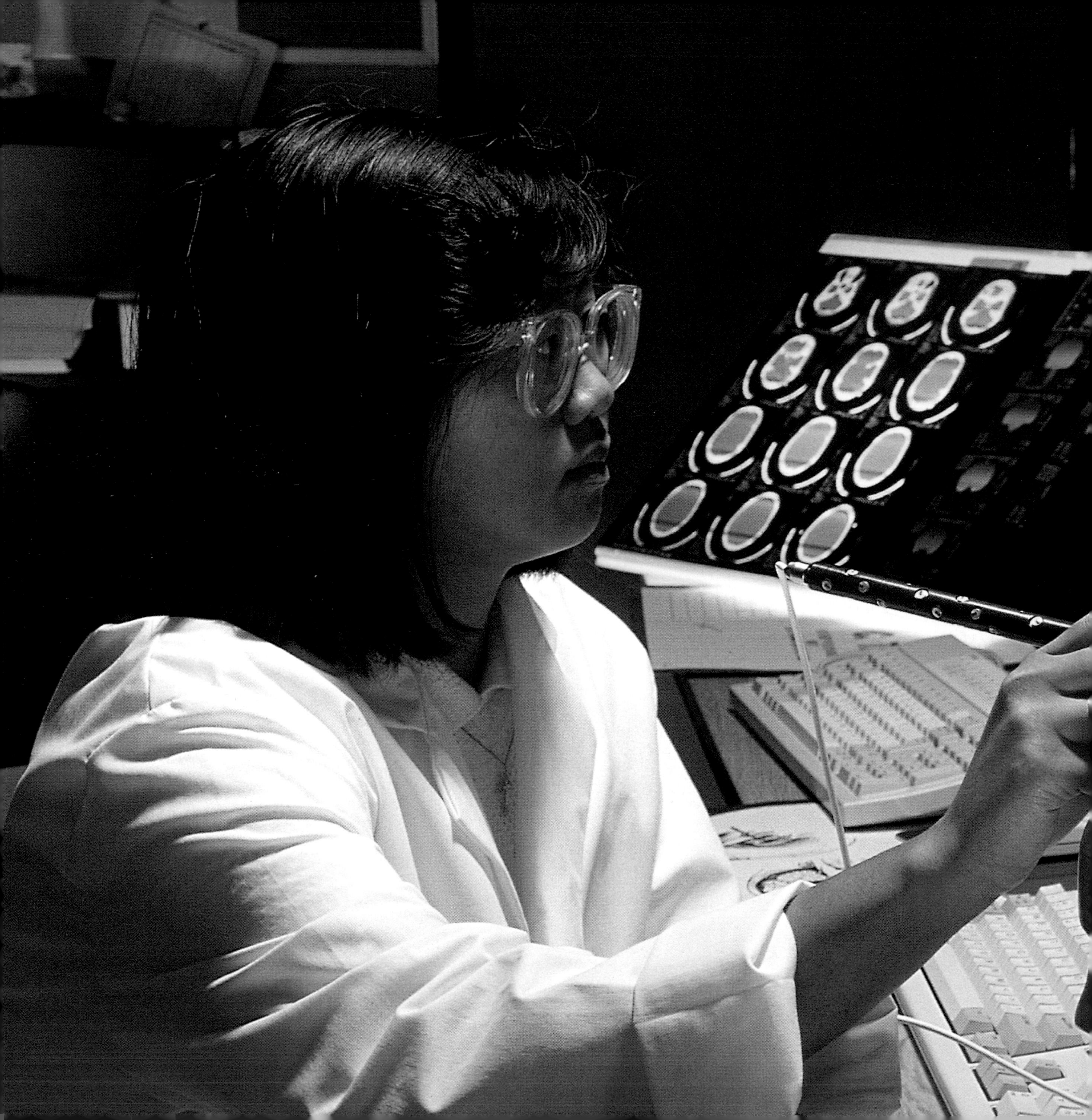

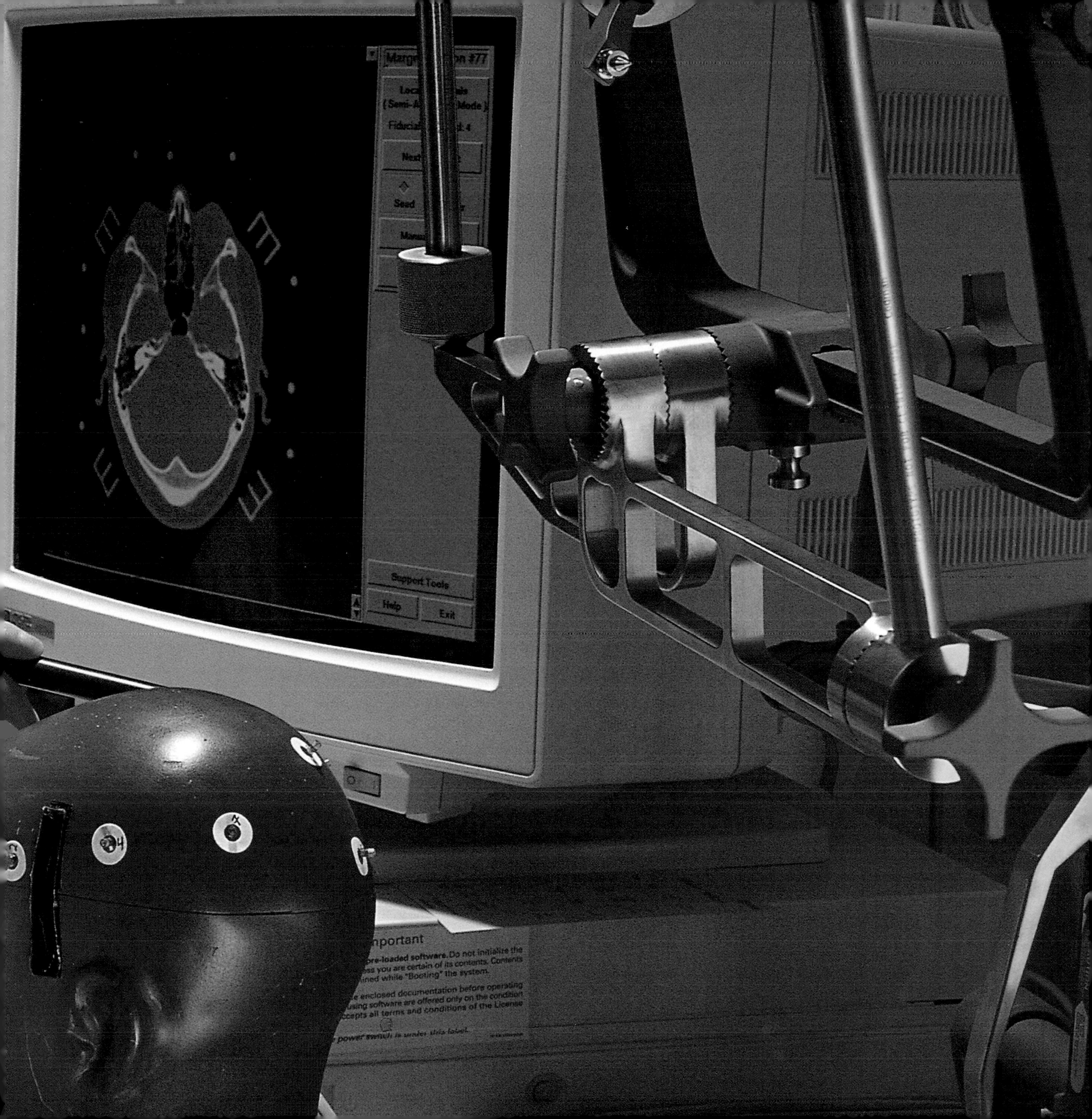

Support Tools
Help
Exit
Seed
Important

RFS/ECM
LEFT
RNG
25
50
100
200
DATA CLR
SAM
A
TSF DATA
RIGHT
TSF
ALARM RESET
AUDIO
F
K
U
DOB
PRTY
ECM
JAM
ERP
BRT

# SAIC
# THE FIRST THIRTY YEARS

EDITED BY STAN BURNS

**TEHABI BOOKS**
DEL MAR, CALIFORNIA

**TEHABI BOOKS**

SAIC: The First Thirty Years was produced by Tehabi Books. *Tehabi*—symbolizing the spirit of teamwork—derives its name from the Hopi Indian tribe of the southwestern United States. As an award-winning book producer, Tehabi works with national and international publishers, corporations, institutions, and nonprofit groups to identify, develop, and implement comprehensive publishing programs. Tehabi Books is located in Del Mar, California.
www.tehabi.com

Chris Capen, *President*
Tom Lewis, *Editorial and Design Director*
Sharon Lewis, *Controller*
Nancy Cash, *Managing Editor*
Andy Lewis, *Senior Art Director*
Sarah Morgans, *Associate Editor*
Mo Latimer, *Editorial Assistant*
Maria Medina, *Administrative Assistant*
Kevin Giontzeneli, *Production Artist*
Curtis Boyer, *Production Artist*
Sam Lewis, *Webmaster*
Ross Eberman, *Director of Custom Publishing*
Tim Connolly, *Sales and Marketing Manager*
Eric Smith, *Marketing Assistant*
Tiffany Smith, *Executive Assistant*
Laurie Gibson, *Copy Editor*
Linda Bannon, *Proofreader*
Ken DellaPenta, *Indexer*

Tehabi Books offers special discounts for bulk purchases for sales promotions or premiums. Specific, large quantity needs can be met with special editions, including personalized covers, excerpts of existing materials, and corporate imprints. For more information, contact Tehabi Books, 1201 Camino Del Mar, Suite 100, Del Mar, CA 92014, (800) 243-7259

Science Applications International Corporation (SAIC) is a diversified high-technology research and engineering company based in San Diego, California. SAIC offers a broad range of expertise in technology development and analysis, computer system development and integration, technical support services, and computer hardware and software products. SAIC scientists and engineers work to solve complex technical problems in the areas of information technology, systems integration, telecommunications, national and international security, health systems and services, transportation, environmental systems and engineering, financial services, and space.

SAIC, 10260 Campus Point Drive, San Diego, California 92121, (858) 826-6000.

www.saic.com

Peggy Walkush, *Project Director*
Key Editorial Advisors and Support:
Sam Carroll
Rose Cofone Schur
Donna Harnden
Ron Knecht
Bill Layson
Dave Poehlman
Judith Shannon
Nola Smith
Matt Tobriner



Library of Congress Cataloging in Publication Data: On file with Librarian at Tehabi books, 1201 Camino Del Mar, Suite 100, Del Mar, CA 92014

**ISBN 1-887656-20-0**

*Photos on previous pages: SAIC's Digicon low-light level sensor flew with the Hubble Space Telescope in April 1990 (pp. 2–3); Neurosurgeons can now see the location of their surgical instruments superimposed on an MRI image of a patient's brain, thanks to a system developed with SAIC expertise (pp. 4–5); Inside a military pilot's helmet, SAIC's "Super Cockpit" technology displays a three-dimensional representation of the real world, as well as essential flight data. For the pilot, this means less chance of error and increased odds for survival (pp. 6–7); Telcordia helps telecommunications providers determine if ADSL technology can help boost data-carrying capabilities of existing telephone lines. (p. 8).*

# Table of Contents

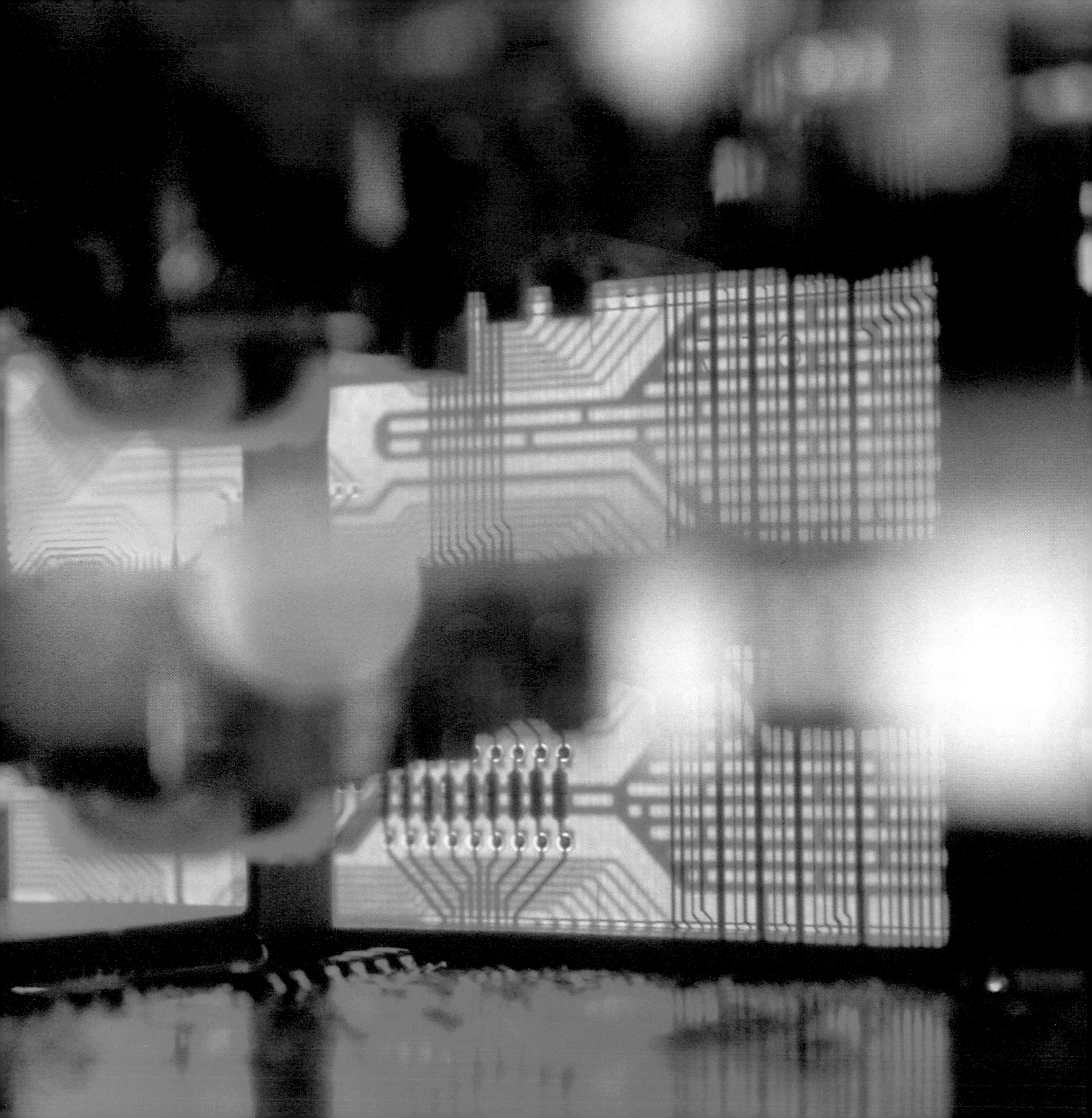

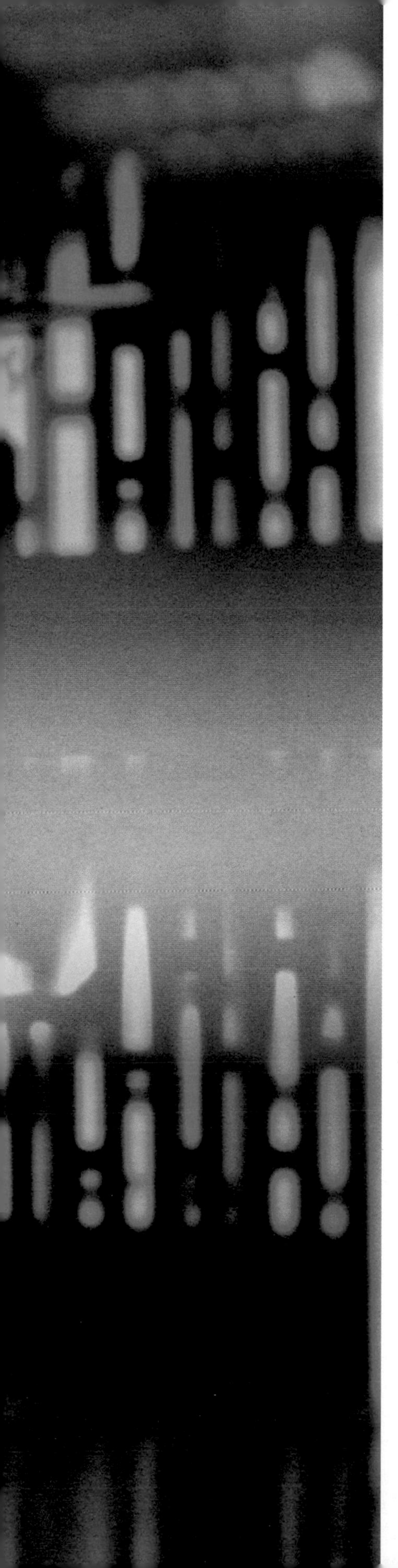

This book captures the story of SAIC's emergence and growth over 30 years, and highlights some of the many technical accomplishments made by some 85,000 past and current SAIC employees. The projects and technologies described in these pages are not all inclusive—SAIC had some 11,000 active contracts in 1999 and countless others in the 30 years prior. Each of these contracts has been significant in contributing to and building upon SAIC's collective body of knowledge and experience.

Our goal has been to create an environment that challenges people intellectually, stimulates them to innovate, provides incentives for them to act entrepreneurially and responsibly, and allows them the freedom to pursue work they are passionate about. SAIC has a tremendous record of accomplishment on services, products, and integrated solutions. It has truly been the intelligence, passion, and entrepreneurial spirit of our employee-owners that have made the Company what it is today. This book is for the employee-owners—SAIC is *their* Company.

—Bob Beyster

SAIC Founder, Chairman, and CEO

# The Birth of a Different Kind of Company

*"The best way I can describe it is 'organized chaos.'"*

—A new SAIC executive on his first Meetings Week experience

*SAIC's quarterly Meetings Week, a long-standing tradition, provides an excellent orientation to the SAIC culture. It is an intense series of technical, management, marketing, and operations meetings that begin at seven in the morning and run until seven in the evening.*

**PRINCIPLES & PRACTICES**

**Being owned by employees allows us to do what we think is best for SAI with a longer-term view and minimizes short-term pressure to compromise our goals and standards. Thus, the customary investor motive of maximizing return on invested capital is replaced by a balanced set of motives combining financial reward, professional accomplishment, and independence.**

*From "Principles & Practices of SAI," written by J. R. Beyster in 1983 to codify the Company's core values in a pamphlet for all employees.*

That's when the formal agenda ends and the informally arranged business dinners with customers or colleagues begin. It is an exhausting week of non-stop data gathering and relationship-building that first-time attendees often liken to "drinking from a fire hose."

Some 1,500 people participate in Meetings Week. Most are SAIC managers, although with manager approval, any employee may attend. A few customers are invited. Perhaps what is most striking to the newcomer is the open, forthright discussions on a wide range of topics—such as company financial performance, potential changes to the incentives system, and contract wins and losses. Attending Meetings Week, one gets a sense of the entrepreneurial spirit that possesses SAIC's employees—a high energy abounds as they greet one another, representing many of the company's 400 offices worldwide.

Because of SAIC's geographic dispersion, diverse business base, and decentralized management, Meetings Week is an important mechanism for SAIC's entrepreneurs to share information on technologies and customers, and to sell their capabilities to one another in order to form partnerships to win large contracts. It is also very representative of the SAIC culture.

On Thursday of Meetings Week, a three-hour meeting called "Management Council" convenes. It was the sole meeting other than the Board of Director's meeting when these quarterly gatherings began in the early 1970s. A summary of presentations is posted on SAIC's intranet following the meeting. SAIC's performance to that point in the fiscal year is discussed with candor, which often comes as a shock to the few customers who are invited to attend.

The April 1999 Meetings Week agenda contained 68 formal meetings addressing subjects as diverse as electronic commerce, counterterrorism, technology commercialization, health business, software management, ethics, Latin American business, and internal training.

At the Management Council meeting on April 9, 1999, shortly after the close of SAIC's fiscal year 1999, Jim Idell, senior vice president for corporate business development, reported that the company's win rate, represented as a percentage of dollars won to dollars bid, climbed to 65 percent in FY(fiscal year)99, up from 62 percent in FY98 and 47 percent in FY97. John Glancy, corporate executive vice president responsible for commercial and international activities, commented that SAIC's commercial revenues reached a record $2.28 billion last fiscal year, exceeding plan by 8 percent and producing 48 percent of the company's totaled revenues. And Duane Andrews, corporate executive vice president, reported that SAIC revenue from its federal government business grew by 14 percent in fiscal year 1999, to more than $2.2 billion in revenue. "This was achieved despite a flat federal budget, which means SAIC successfully captured a larger market share," said Andrews.

SAIC Founder, Chairman, and Chief Executive Officer J. Robert (Bob) Beyster.

In the mid-1970s, the employees presented Bob Beyster with this cartoon, below right, depicting Beyster attempting to steer the Company's fiercely independent "entrepreneurs" in one (the right) direction. In 1998, SAIC's employees being photographed for the cover of Forbes, below left, convinced the photographer to capture a similar scenario.

Dr. J. Robert Beyster, founder, chairman, and chief executive officer of SAIC, and traditionally the final presenter at Management Council, offered his observations of the company's performance and areas that needed focused attention.

"For the year just ended, revenues were $4.7 billion. That was $400 million over plan." Beyster paused and looked out at the audience of 1,000, primarily the managers who run the company. "Net income was $150 million, 24 percent over plan. We have $1 billion in the bank and our stock price increased 78 percent this past year." The audience broke into applause at the success they had achieved together. SAIC was their company, literally, and one of the largest employee-owned companies in the United States.

Beyster's tone quickly shifted from satisfaction to warning. "We work better when we're scared," he said. He began to list the concerns he saw hovering on the horizon. The hands-on scientist who started this company in 1969 was determined that no one would drift into a world of self-satisfaction and jeopardize what was still being built. The non-traditional company he founded in 1969 with one contract and rented space in La Jolla had moved into the ranks of the Fortune 500 list 30 years later—the fast-paced growth still being fueled by the shared values that define SAIC and distinguish it from competitors.

# Bob Beyster: A Different Kind of Builder

Below, left, a young Bob Beyster surrounded by two friends on the family Buick in Grosse Ile, Michigan. (Below right) Betty and Bob Beyster, with their children Mark, Jim, and Mary Anne (left to right) standing behind them. Bob Beyster's photo, far right, from the University of Michigan Navy V12 College Program in 1944.

"I'm not the brilliant flash-of-inspiration type entrepreneur . . . I'm more of a persistent, workaholic type. I get going in a certain direction and suddenly things start happening—a builder type, basically . . ." (Bob Beyster's description of himself in a speech delivered in 1988).

In addition to always being a workaholic, Beyster developed intense loyalties early in life. He grew up in a traditional Midwestern family in the early 1930s expecting that he would have a career like his father as an engineer at General Motors. He had no entrepreneurial dreams of starting his own company and thought that Detroit and General Motors would be his home all his life. During the Depression, having a good-paying, secure job was important, and wild dreams of a new and different type of company in faraway California did not enter young Beyster's head. Then World War II hit as he was finishing high school, and he joined the Navy.

It was the Navy—rather than his parents or his own dreams—that started him in a serious scientific direction.

The Navy did basic aptitude testing and decided to send Beyster to the University of Michigan to study mathematics and

physics. About the time the war ended, he graduated from Michigan with an engineering degree. He completed his tour of duty in the Navy in 1946.

Having been detoured from his expected route to General Motors as an engineer, Beyster took another round of aptitude tests after leaving the Navy. "They (the testers) said 'whatever you do, don't become an attorney,' said Beyster many years later. "Many times I have thought how right they were." He was told that the Navy had made the right decision, and he should go back to school for further studies in mathematics and physics. Beyster returned to the University of Michigan, where he received a Master's degree in physics in 1947 and a Ph.D. degree in nuclear physics in 1950.

General Motors was now a distant memory, and because of the education he received courtesy of the Navy, Beyster was headed in a direction that would rely heavily on his physics background. After a short stint at Westinghouse, he worked for five years at the Los Alamos National Laboratory, honing his abilities as an experimental physicist and working on important scientific problems. There he conducted research and co-authored papers with Nobel Laureate Hans Bethe, one of the famous Manhattan Project scientists.

"I like doing experimental work," Beyster added in the 1988 speech as he described his career. "I could have been perfectly happy staying at Los Alamos and doing research forever . . . But many people at the lab were leaving and coming to a place called General Atomic . . . And my wife wasn't too happy at Los Alamos, which had sort of a profound impact on me."

Now a nationally recognized authority in nuclear weapons effects and nuclear reactor physics, Beyster was hired by General Atomic to build and run an accelerator physics department. He assembled the facility, grew it to a staff of 130 people, and made it profitable. He worked there for twelve years, and by the late 1960s, General Atomic had been absorbed by Gulf Oil and was no longer interested in the work performed by researchers like Beyster. For the first time in his life, he felt the need to be in charge of his own destiny.

"I have a strong belief," said Beyster in 1988, "that people should work for organizations in which those above you, particularly at the top of the organization, are interested in what you're doing."

He then left General Atomic and started Science Applications, Inc. (SAI) without, as he says, a "grandiose plan for the future." He expected the new company to be reasonably profitable, but he did not start SAI in order to generate large financial rewards for himself. His immediate goal was to create "a good place where I could work and maybe a few people could join me."

At Los Alamos and at General Atomic, Bob Beyster worked with many scientists who were eager to capitalize on their knowledge and skill. "These were guys who would work at the lab for a while, get an idea, and ask the lab if they cared if they took the idea and went off and started a company. That was always very distasteful to me," said Beyster. But eventually the environment changed at General Atomic and Beyster decided to leave in order to create a more comfortable place to do the work he loved.

## How It All Began

Beyster's highly regarded scientific credentials played a part in his getting the first contract for his new company: a $70,000 contract with the Defense Atomic Support Agency to study the output from nuclear weapons. From the very beginning, the new company was working on complex scientific issues that had an important national impact.

In the 1960s, atmospheric testing of nuclear weapons had been stopped, and by the time Beyster formed his company, he was using accelerators to simulate nuclear results. He was more an experimenter with, than a developer of, accelerators, and he used them to help refine man's knowledge of nuclear weapons. The first contract for the new company laid important groundwork for other contracts in the areas of nuclear blast effects and nuclear effects simulation.

With a contract in place, Beyster and his first employee, computer expert Don Huffman, sat down and made a list of potential names for the fledgling company. They chose Science Applications, Inc. (SAI) because it seemed to cover everything the company would be doing, and its simplicity set it apart from the fancy names his former colleagues used for their new companies, names that people either couldn't remember or names they laughed at. (In 1984, SAI was merged into Science Applications International Corporation, SAIC, indicating that the company had broadened beyond national boundaries.)

Beyster became an entrepreneur at the age of 45 by accident. He became an entrepreneur in order to do the type of work he wanted to do when he could no longer do that work in a larger corporate setting. His experience at General Atomic was an important preliminary step. There he had sharpened his scientific skills, he had learned what he liked to do, and he had managed people in a setting where priorities shifted quickly and resources were not easy to come by. He had learned pragmatic skills, and SAI was built on a simple formula: hire very smart people, give them the freedom to build business in areas they are passionate about, and reward them for their contributions with ownership of the company.

*"I was not sure that the company would ever amount to anything, but if it did, I wanted to share it . . ."*

The employee ownership idea was an intuitive response to a practical question. When Beyster incorporated the new company, he was asked who would own it. His quick response was that it was only fair that the people who built it should own it. As he hired new people, he told them if they brought in new contracts, they could buy stock and become owners of the Company. By the end of the first year, Beyster had diluted his ownership to 10 percent, and there were 16 scientific people working at SAI. Years later, Beyster described his early feelings: "I was not sure that the company would ever amount to anything, but if it did, I wanted to share it fairly with those who made it worth something."

In addition to its nuclear weapons work during its first year, SAI was also hired to study nuclear safeguards for the Atomic Energy Commission and to work on a radiation-based system of cancer therapy for the Los Alamos National Laboratory. For the first year, revenues totaled $243,000 and the Company was profitable.

In 1969, Betty Beyster gave her husband Bob this poster, bottom right, for his new office. "It seemed to express how the small organization was working together," Mrs. Beyster said. "There were only four or five employees at that time, and it described the support they gave each other in those early years. I'm sure that none of them realized at that time what the Company would grow to be." Many years later, Bob Beyster said, "At first I didn't realize how true it really was. We had people who were willing to try new ideas to solve difficult technical, marketing, and management problems, people who would give it a try and not put all the issues in the 'too hard' category. We became a 'can do' company. I have thanked Betty many times for that picture." SAI's first office, left, in La Jolla, California, came cheap—$2.40 per square foot per year—not bad for an office with an ocean view.

The young, growing company settled into offices on Prospect Street in La Jolla, below, and continued to work on an array of scientific products like The ATP Photometer, below left, which measured living organisms in liquid samples, and the Moisture Meter, right. In 1974, the Company installed a DEC-10 computer, below right, to expand its capability for systems analyses and scientific investigations.

## Building the Business

As the new company began its second year of operations, SAI's plan called for increasing revenues to four times the level of the first year and boosting profitability. The Company hired new people at the rate of several a month and put them to work in areas of their own interest in both government and private sectors. Finding new work in the commercial sector was identified as a priority, both to enhance profitability and to use the Company's base of technology as broadly as possible. In 1970, SAI's expertise in radiation effects and radiation treatment was first applied to health care.

By the second year, all of the employees shared a common bond since the results of one were seen by all. A competitive spirit, together with their ethical and professional principles, motivated SAI employees to perform at a superior level—there was no room for poor performance. A self-selection process began to occur, and those not able to function well in that environment left and those who were comfortable with the high standards of performance did well.

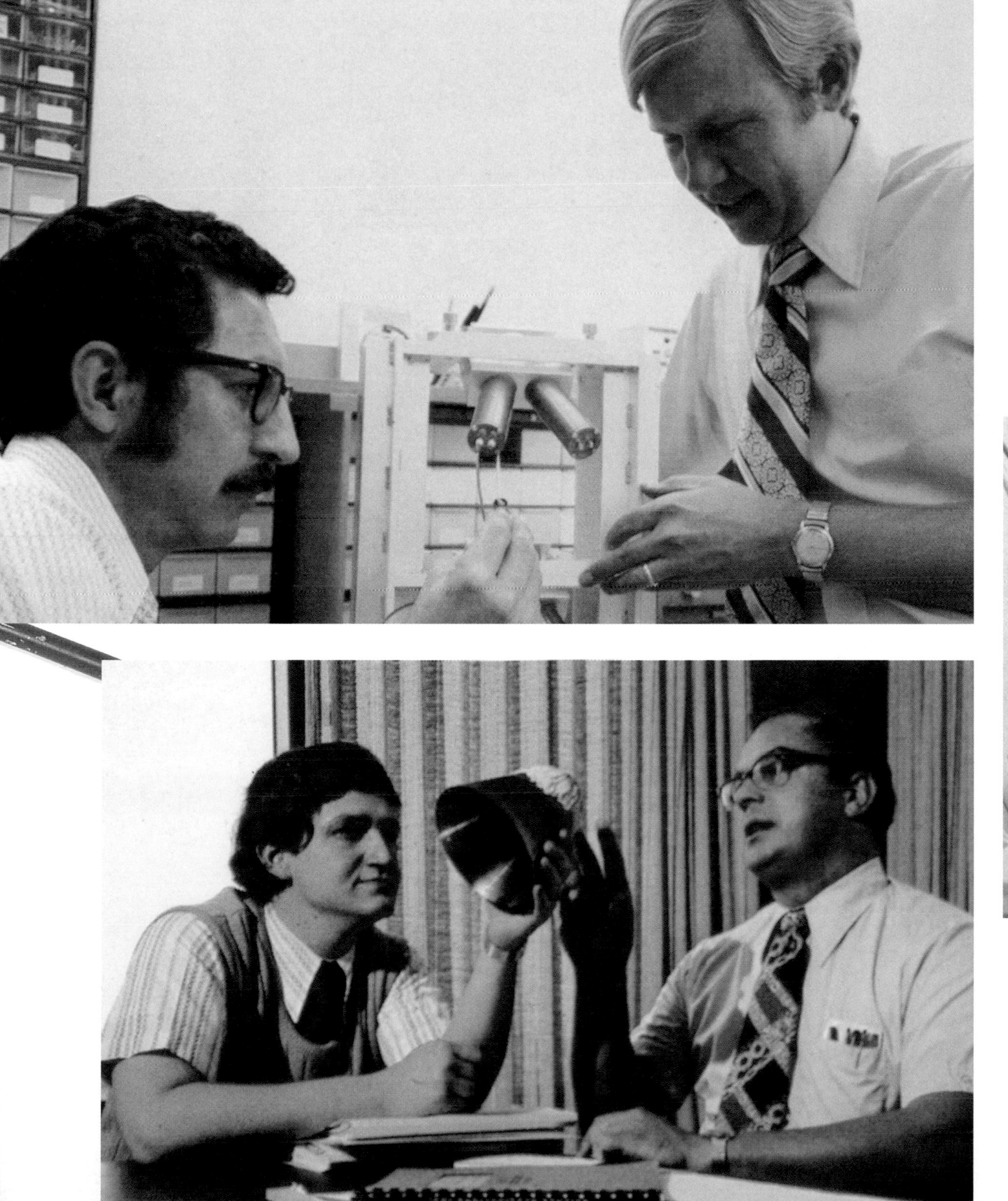

In the early 1970s, the Company was a provider of high-technology research services to the federal Government. Better composite materials for nose cones, below left, improved detection of neutrons, top left, and the precise measurement of X-rays, below right, were a few of the Company's many areas of research and development.

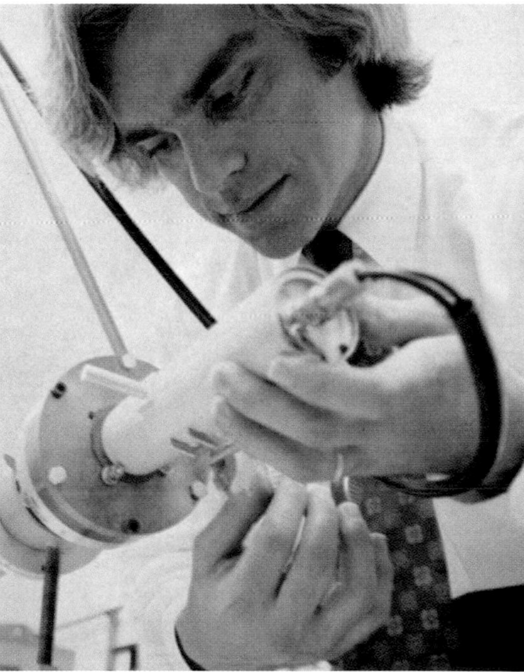

Long before electronic security and computer keyboards, SAI used prevailing techniques and technology. The Company's work has always required security, and early badges, right (of longtime employees Bill Wright and Amanda Stokes), provided limited access to Company facilities. This typewriter, below right, was the Company's first, and important early contract proposals were pounded out on it. Communications were important early on, and the Newsgram, far right, was a publication for employees. The folder from the Company's JRB subsidiary contained a marketing brochure for customers.

# Newsgram

**WEATHER REPORT**

La Jolla
Partly cloudy with some fog in a.m. Ray of sunshine in midday. Partly cloudy with some fog in p.m. High in the 80s on the coast.

Washington, D.C.
Cloudy, windy, rain. High in upper 30s.

LA JOLLA, CALIFORNIA DECEMBER, 1977 A PUBLICATION BY AND FOR SAI EMPLOYEES

## News in Brief

Frank C. Alexander has been named SAI's Senior Vice President for Administration & Chief Administrative Officer. Frank was fficially installed at the September 15, 1977 oard of Directors Meeting.

gala open house was held to announce the ening of the JRB Architects, Inc., St. uis, MO (a JRB subsidiary).

has wit, charm, personality, and a g desire to increase company-wide com- cations and who happen to be two of most gregarious staff members? The Newsgram Staff!—Jessica van Andel herry Gibson. (See page 19 for more on the staff and "volunteer" editor, Pearson).

### Special Achievement

kewes-Cox has been working as a staff scientist at our Pleasanton about a year and a half, during he also was pursuing a double major at UC-Berkeley in physics and applied mathematics. Peter recently has been awarded the 1977 University medal as the most distinguished graduating senior at UC-Berkeley, having completed four years of study with 23 A pluses, minuses. He also served as a Delta Phi fraternity, as undergraduates, mastering German, and Latin, and submitting articles for publication in the *Journal of Physical Chemistry*. In his work at SAI, Peter has been involved with computational physics, partially ionized X-ray filters, thermodynamic equilibrium tropospheric photochemical mechanisms, methane kinetics and energetics, and oil shale retort modeling. Peter is now a full-time member of our staff. Congratulations to Peter!

## Goals, Objectives and Reasons for the SAI NewsGram

As SAI approaches the end of its eighth successful year, we have grown to a rather large "small company" of 1997 employees. We have grown from a single office to approximately seventy facilities spread geographically across the country. In the past, several of our offices have produced their own newsletters, the most widely circulated being the SAI NewsGram, published quarterly in La Jolla.

This new monthly version of the SAI NewsGram is dedicated to company-wide communication, with the ultimate goal of providing relevant information to all echelons of SAI, JRB, and their subsidiaries. The intent is that more effective communication will allow better decision-making at all levels, thus supporting SAI's belief in individual responsibility for professional performance, satisfaction, and growth.

It is our desire to help people know and understand what is occuring in the company, and to be aware of our plans for the future. As shown in this issue, we will utilize a basic format that covers the following areas:

**Offices/Locations** — Individual offices, labs and groups will be featured monthly, detailing facilities, staff and areas of expertise. A wide base is planned, covering both administrative and technical groups throughout the organization to facilitate a better understanding of our company.

**Featured Groups** — The first eleven issues will cover the major management groups with a description of the organization, management, administrative group and areas of endeavor.

**Who's Who/Promotions** — This section will recognize employee advancements within the company, introduce new employees and feature key staff members.

**How To/What's Available** — This section will answer questions and explain procedures for obtaining information already available within SAI. Subject matter will range from questions and answers to in-depth articles.

**Contracts** — Each issue will feature a current list of new contracts by division for the previous month. In addition, specific contracts will be highlighted with emphasis ranging from our small, unique efforts to our larger, long-range efforts.

**Special Achievements** — We plan to give monthly coverage of individual and group achievements. Noteworthy personal achievements will also be featured.

**Calendar** — In addition to the monthly calendar of events, reports of the past month's activities of interest will be detailed. Our plan is to set up a working system for informing all interested employees of company-sponsored activities to ensure maximum information and allow participation.

**Jobs Available** — Each NewsGram will set aside space for a listing of jobs available within the company. These "Want Ads" will cover all areas of available employment—clerical, administrative and technical—with a brief description and contact. By this method, we hope to keep the staff advised of opportunities for growth and/or change within the company.

The success of our objective—improved communication vertically and horizontally throughout the organization—depends on you, the individual employee. Without your interest, support and input we will not achieve our goal. We welcome and encourage all comments and suggestions as we strive to continually improve on this original premise. ■

**INCREASE COOPERATION THROUGH COMMUNICATION VS. COMPETITION IN UNCERTAINTY**

Because of the strong emphasis on bringing in new business, the majority of the Company's employees have always been involved with customers on a daily basis. In the early years, the person who wrote the proposal for the customer was also the person who did the work called for in the contract, and that same person managed the financial aspects of the contract. Since employees were constantly in the marketplace talking to customers and prospective customers—not spending all of their time in the laboratory—they could respond to changes affecting their customers and could create scientific solutions that were relevant for changing conditions.

By its third year, the Company had a new office in Huntsville, Alabama. The office captured the largest contract in the United States for independent software verification and validation This computer systems contract supported the U.S. Army's Safeguard anti-ballistic missile, and it put the Company on the map in the defense industry.

The planetary sciences group furnished important analyses for the planetary missions of the U.S. Space Program, below. Knowledge gained from these explorations was used to enhance the effectiveness of those that followed.

In 1973, SAI received more than 100 contracts from various parts of the Department of Defense. During the following few years, military computer systems became the Company's fastest-growing business. The expertise of many specialists from various disciplines was the critical factor in SAI's ability to undertake numerous government contracts and to achieve recognized status within the research and development community. The Company became increasingly involved in long-range military planning efforts that positioned SAI scientists at the leading edge of military technology and thought, laying a foundation for a wide variety of future military projects.

In 1973, SAI's revenues hit $21 million, more than double the year before. The breadth of the Company's capability was well established, employees totaled 703, and SAI was vigorously profitable with net income of $695,000. The Company's formula for success was taking shape.

SAI quickly began using knowledge and skill gained from national security work to undertake projects for the nuclear energy industry, left, and health care, below left, which would become a vital part of SAI's future. Larry Kull, shown on the right in the photo at left, started with SAI as a physicist in 1970 and worked his way up to the position of president and chief operating officer before retiring in 1996. The chromosome analyzer, below right, was developed for clinical analysis in cytology and genetics laboratories.

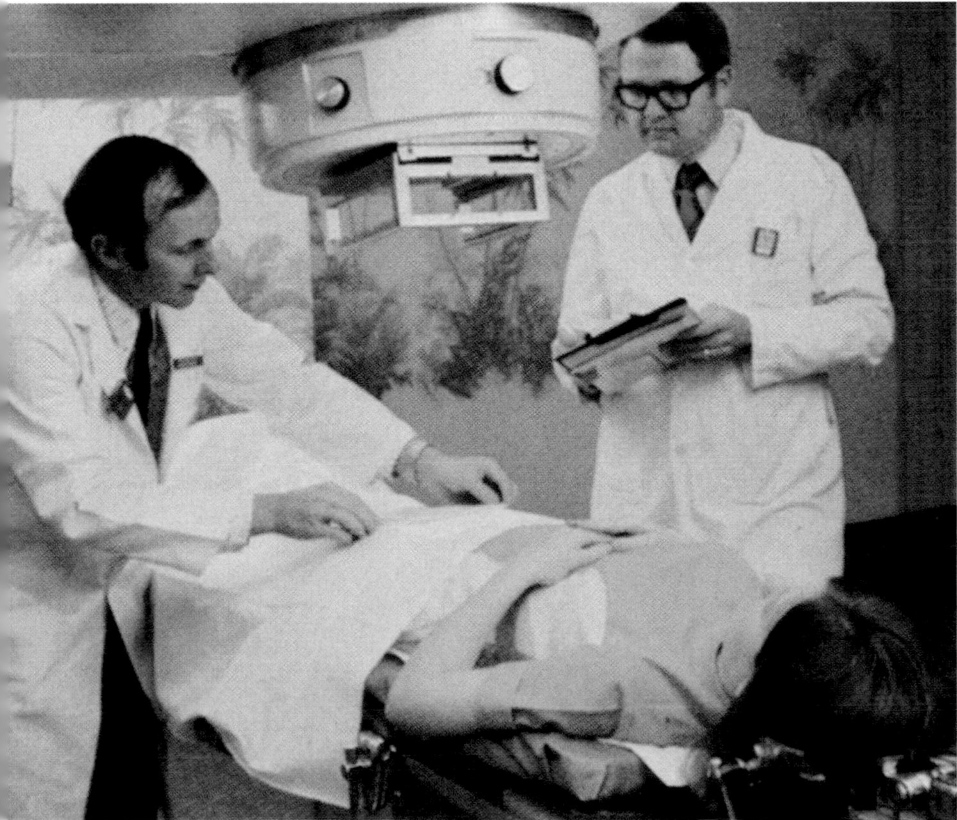

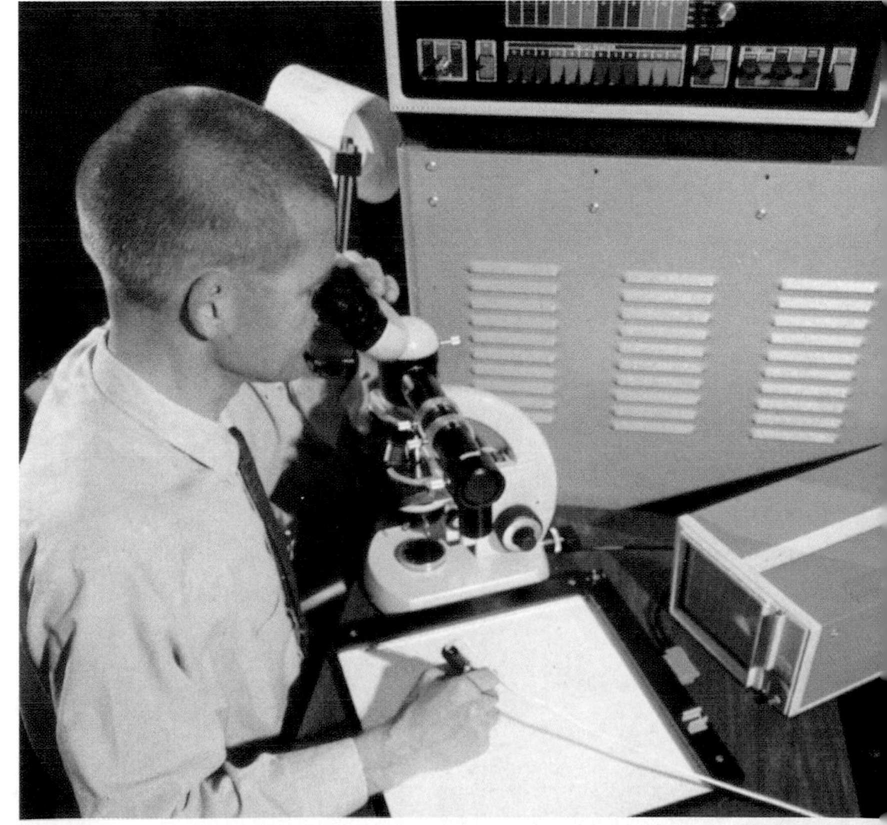

# Going National: A Foot in the Door in Washington

SAI's early expansion into the Washington, D.C., market reinforced the Company's commitment to research and development in support of U.S. national security programs. Bill Layson, center in the photo at right, opened the Washington-area office in 1970. Below, as part of an effort to study particle fallout from nuclear explosions, SAI collected dust samples from volcanic eruptions and TNT explosions by driving down wind in a pickup truck to gather dust in the bed of the truck. This bottle contains about a teaspoon of dust.

During SAI's second year, the Company discovered a new way to grow: establish offices near major customers and hire the best possible people to meet the needs of that customer.

Bill Layson opened SAI's first office outside of La Jolla, close to Washington, D.C., in September, 1970, staffed by himself and a secretary. The Washington office was a short distance from the Pentagon and numerous defense agencies, and Layson spent the weekend before the doors opened writing a proposal for a defense contract. The office won that contract—it was for several hundred thousand dollars and ran for a year—and SAI's de-centralized operating strategy was off to a successful start.

Being close to the customer quickly became a crucial piece of SAI's formula for success, and a decentralized management structure evolved to provide SAI's "entrepreneurs" the freedom to build their own businesses. SAI attracted national experts to the

Company because of the freedom it offered—including allowing them to locate wherever they desired as long as they could build a successful business near a customer.

The SAI office in Washington was an important step in establishing a remote facility close to important government customers. The mission of the new office was to focus on space and missile systems work generated by the nearby federal agencies and to identify possibilities for diversifying the Company's business.

A deal was struck such that if the new Washington office were able to generate $1 million in revenues in its first eighteen months, all of its employees would receive stock options. Soon the Washington office was overflowing with people and the $1 million goal achieved. In the usual SAI style of keeping costs down, used furniture was everywhere in the offices and in the halls. Business was steady, and the early national security work expanded to include health care contracts with the National Cancer Institute.

Nearly thirty years later, the Company had over 13,000 employees in the metropolitan Washington area, and national security contracts were one part of its widely diversified base of business.

Bill Layson retired from the Company in 1999 after twenty-nine years of service as a key executive, Board member, and founder of the Incentives Committee.

This capabilities brochure from 1973 was one of the earliest color marketing pieces the Company produced. Support of national defense was to become the Company's mainstay, and included scientific breakthroughs in underground nuclear test experiments on energy coupling and electronic survivability.

An early part of the SAI philosophy was staying close to the customer. The Company's Huntsville, Alabama, staff continued the SAI focus on high-quality technical achievements. John McRary (center of photo) managed and grew the Huntsville office over 20 years, served as a senior executive for many years, and remained as a Board member in 1999.

Many SAI employees were experimental physicists who were open to all sorts of ideas and willing to try novel solutions to meet the customers' needs. Changing needs were generally seen as a challenge rather than an impediment. With a large amount of decentralization and a horizontal, spread-out organizational structure, entrepreneurial scientists working at SAI felt like they were running their own small companies.

With the 1973 energy crisis, the Company's energy research expanded rapidly. Ed Straker (right) built much of the Company's energy business, and in 1999 led the Integrated Solutions Sector, which generated $270 million of the Company's revenues.

SAI analyses and experiments helped the U.S. military understand the effects of nuclear weaponry. The Company developed simulators and nonexplosive techniques to evaluate the thermal output of nuclear devices, this page and at right.

The Company performed a substantial amount of research and development in nuclear weapons technology. SAI's first contract in 1969 with the Defense Atomic Support Agency involved the study of the neutron output from nuclear weapons. In subsequent years, SAIC performed numerous studies of different aspects of the output of nuclear weapons and the effects of that output. SAIC's work on nuclear weapons effects dealt with virtually every aspect of the weapon other than its actual construction.

This nuclear weapons' effects work was an important technology niche for the Company, one that continued to grow for many years. Since atmospheric testing of nuclear weaponry had long since stopped, SAIC used simulations to study the technical definition and impacts of nuclear dust clouds. Mathematical constructs and computer programs were developed to assess the impact of the dust clouds on military systems. The analysis of older data from atmospheric tests presented challenging research problems in confronting a real and present concern. High explosive simulations, computer simulations, and the study of volcanic eruptions provided crucial research that required constant updating as new and different missiles, weapons, and applications were designed and built.

**PRINCIPLES & PRACTICES**

**In essence, we expect our employees to treat each other in the same manner in which they would like to be treated. SAI absolutely requires of its employees basic honesty in representing their work correctly, in any and all financial acts within the company, and in representing their individual credentials and past accomplishments. The integrity of the company is exemplified through the actions of all employees. Professional integrity is essential to fulfill contractual obligations, to maintain the quality of our products, and to uphold the reputation of SAI.**

*From "Principles & Practices of SAI," written by J.R. Beyster in 1983 to codify the Company's core values in a pamphlet for all employees.*

The SAI team grew at a rapid pace during the 1970s as additional scientists, technicians, and administrative staff were hired to keep pace with the business the Company was generating. By 1975, the Company had over 1,000 employees and net income exceeded $800,000. In 1974, a new research area focusing on electro-optics started at a new location in Ann Arbor, Michigan, right, as part of the Company's Military Sciences Group.

Understanding this nuclear technology was vitally important for the Government because of the huge financial ramifications. Decisions had to be made for allocating funds for military hardware, and engineering problems for the latest aircraft and missiles needed to be studied thoroughly and correctly as those funding decisions were made. SAIC continued this work until the end of the Cold War in 1990. By the time the military needs for this research had declined, the Company had accumulated a vast body of knowledge and expertise that was transferable into other areas of nuclear research.

The Delta rocket, left, carried the SCATHA satellite into space. SCATHA carried about a dozen experiments to gather data to analyze the phenomena of spacecraft electrical charging. SAI provided the on-site real-time operations planning at the satellite control facility for the first six months of the mission in 1979. As space exploration became a larger concern for the Air Force and NASA, the Company's research projects expanded to cover both military and non-military aspects of space.

The Company's energy and environment work in the 1970s required an expanded staff and increased research activities. Air pollution was a national concern, and SAI developed gas analyzers that measured levels of air pollutants.

Energy and environmental issues were important national priorities in the early 1970s, and SAI worked on projects ranging from energy conservation to the development of new energy sources. The Company used the base of knowledge it acquired in its first military contracts, which studied radiation effects, to undertake detailed risk assessments for the nuclear power industry and federal regulatory agencies. These projects involved a substantial transfer of technology from the governmental defense arena into the world of nuclear energy. The Company would later become the leading supplier of probabilistic risk assessments for the U.S. nuclear power industry, then branch out to serve international customers. By 1997, when damage occurred to the nuclear power plant at Three Mile Island, SAI had developed a highly renowned team of experts who were mobilized to work on stabilizing that plant.

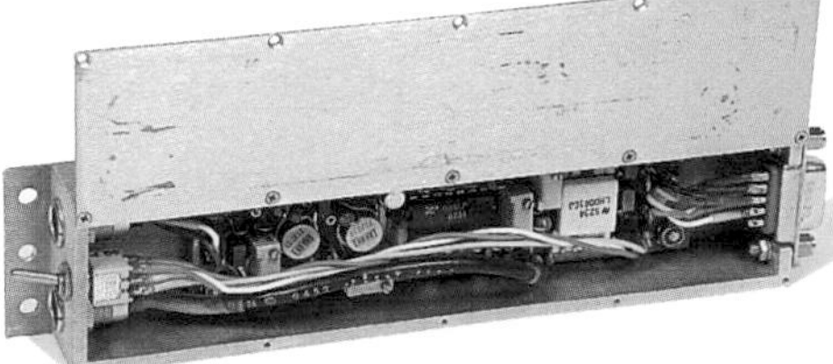

As part of the effort to make the United States more energy self-sufficient, SAI undertook a major study of the Southern California continental shelf in anticipation of an expansion of oil and gas exploration and production. Another study of the economic and environmental aspects of underground petroleum storage expanded the range of national energy choices, aiding the U.S. Government in directing its national energy policy.

By 1975, SAI employed 1,213 people in 45 offices throughout the United States. Eighty-seven percent of those employees had degrees in physical sciences, engineering, and mathematics. Fifty-eight percent of the employees had Master's or Ph.D. degrees. The same percentage had over ten years of experience. Nearly half of the employees were shareholders of the Company. Revenues in 1975 passed $45 million, and the challenge for SAI's leaders was to maintain the working environment

In 1974, SAI's Board of Directors was reformulated to balance insiders with distinguished outsiders. The SAI 1974 Board members were, front from left to right, W. E. Zisch, J. R. Beyster, B. J. Shillito, standing from left to right, W. M. Layson, L. C. Fricker, and T. F. Walkowicz. Several became long-term key advisors to the Company. In 1999, Barry Shillito remained a Board member Emeritus. Bill Zisch held Board member Emeritus status until his death in 1998. Bill Layson retired from SAI in 1999.

**PRINCIPLES & PRACTICES**

**SAI is a company for professional people who want to perform superior scientific and technical work; who are willing to work hard to do it; who want to have a say in the policies and management of the company and feel that the company is their company; who want to be exposed to a minimal number of distracting outside influences and pressures; and who want to be fairly rewarded for doing good work both from a recognition standpoint and from a financial standpoint.**

*From "Principles & Practices of SAI," written by J.R. Beyster in 1983 to codify the Company's core values in a pamphlet for all employees.*

they had created and sustain the Company's rapid growth. Their intuition about building a diverse company had worked through the early years—the amount and type of business had grown, the culture of passionate scientists doing important work had taken hold, and the novel concept of employee ownership was giving large numbers of people a chance to own part of the Company.

Just prior to the close of the 1970s, SAI won a U.S. Navy contract to study the feasibility of integrating a command, control, and communications center for the Saudi Arabian Navy. That work continued for several years and would culminate in the win of a several-hundred-million-dollar procurement in 1985 to build a turnkey facility for the Saudi Navy. This effort would later be regarded by the Company as its single most defining moment of the 1970s—launching SAI as a major systems integrator and setting into motion a customer relationship that remained in place in 1999.

SAI's win of a contract to study the feasibility of integrating a command, control, and communications center for the Saudi Navy would later lead to a major procurement to build the center, this page. In 1973, SAI installed its own computer center, left, in Huntsville, Alabama.

At the end of 1975, SAI had 559 shareholders compared with 408 the preceding year. The Company continued to actively involve its employees in Company ownership and to provide continuing programs to nurture their professional capabilities. The practice of sharing equity and empowering employees to perform at high levels of achievement were parts of the SAI philosophy that were well-entrenched as the Company began to grow rapidly during the 1970s. After almost twenty years, the phenomenal success of employee ownership was recognized by *Fortune* magazine, right.

# Taking Stock: Employee Ownership Makes the Difference

"Those who contribute to the Company should own it and ownership should be commensurate with a person's contribution and performance as much as feasible."

That principle has provided the foundation on which SAI's stock ownership system was built and on which it continues to evolve today. The concept of merit-based ownership led to what Beyster now calls "entrepreneurial employee ownership" to differentiate SAI's program from entitlement programs in other companies in which employee ownership is based on a percentage of an employee's salary.

In the early years, stock ownership was a privilege earned by those who generated new business for the Company. Profitable contracts were the lifeblood of the young company, and the people who brought in business were rewarded with the right to buy stock. "I had to borrow $500 to buy my first stock," one employee recalled, emphasizing that stock ownership was a privilege, not a gift.

By 1973, some SAI employees were concerned that the SAI stock they owned could not be readily sold and felt that their interests would be better

served if the Company went public so they could sell their stock. SAI's leaders felt that the advantages of being a private company continued to be important, so they created a mechanism for employees to buy and sell stock. In 1973, SAI created Bull, Inc. as a wholly owned broker-dealer subsidiary registered with the Securities and Exchange Commission to facilitate the buying and selling of SAI stock among employees. It was a unique approach then, and to this day no internal market exists as sophisticated as Bull, Inc. Approximately 45 people participated in the first trade in early 1974. More than 3,500 participated in the April 1999 trade.

Until 1981, employees who left the Company were permitted to continue to hold their shares of SAI. Since departing employees often kept their stock, the Company's employee ownership was gradually diluted. In 1981, the Company established the right to repurchase stock at the termination of a person's employment to ensure that the Company be predominately owned by those currently contributing to its success.

MAY 4, 1992

FORTUNE

SCIENCE APPLICATIONS INTERNATIONAL

# TALK ABOUT PAY FOR PERFORMANCE!

**SCIENCE APPLICATIONS INTL.**

STOCK PRICE
Near end of quarter

1/10/92 $11.15

SAIC's stock is not publicly traded. Its price is tied to a performance-based formula that reflects changes in net income and its competitors' P/Es.

| | |
|---|---|
| SALES (latest four quarters) Change from year earlier | $1.3 billion Up 10.5% |
| NET PROFIT | $33.6 million Up 9.5% |
| RETURN ON EQUITY | 9.2% |
| TOTAL RETURN TO INVESTORS 9/12/86–1/10/92 (annual rate) | 10.2% |
| PRICE/EARNINGS MULTIPLE | 14.9 |
| DIVIDEND YIELD | None |

IS BRILLIANT management the secret of success at Science Applications International Corp., a thriving $1.3-billion-a-year high-tech research and engineering company in San Diego? "No," chuckles J. Robert Beyster, the company's founder and a nuclear physicist by training. "It's certainly not *that*." Then what is it? Employee ownership, he maintains. Workers and directors own 46% of SAIC's shares outright, while 41% are reserved for the company's ESOP and 401(k) plans. "If there is a problem with a contract that could make the stock go down," says Beyster, "employees sound off about it. The fact they are concerned puts teeth into management."

Some teeth. For the past two decades, SAIC's stock price has climbed at a 27% annual compound rate, though the pace has slowed in recent years. It's almost impossible to get hold of any of this private company's paper. The only shares held by "outsiders" belong to consultants, former employees, and ex-spouses. Once every quarter, the privileged can buy and sell stock through Bull Inc., a broker-dealer subsidiary of SAIC that's gotten a green light from the Securities and Exchange Commission. To preclude any hint of manipulation, the company ties its share price to a set of financial criteria—among them, changes in net income and stockholder equity. It also calls in an outside appraiser to vet its decisions.

What's rare is that SAIC doesn't limit the goodies to an in-group at the top. Nearly 7,000 of its 13,500 employees own shares outside the retirement plan. Nor is this stock granted simply as an entitlement, as is typical under most broad employee stock ownership schemes. Founder Beyster, who owns just 2.5% of the equity, insists on linking ownership to performance, measured by, among other things, sales targets or how well a manager handles contracts. (To maintain that nexus, only employees recommended by their managers can buy stock on SAIC's quarterly trading days, though anybody can sell.) The awards add up: A production technician with SAIC for 14 years recently retired with $300,000 in stock.

Tying employee welfare to corporate results has overcome institutional lethargy and helped make SAIC surprisingly nimble. While many defense contractors struggle to diversify, SAIC has made the move with grace and speed. Just six years ago, more than two-thirds of its sales came from providing technology, analysis, and engineering support to the Pentagon for Star Wars, arms control verification, underwater surveillance systems, electronic warfare, and the like. While that business continues to grow, defense now accounts for just over half of revenues. SAIC's big new growth markets: health care and pollution prevention and cleanup.

The same geologists, hydrologists, and other environmental experts who once contemplated the chilling prospects of a nuclear winter are now working for NASA on the problems of global warming. Others are developing innovative technologies to help the EPA clean up the worst of the country's hazardous waste sites. One of the most promising—using bacteria that gobble up waste—was employed on two-thirds of the beaches of Alaska's Prince William Sound that were polluted by the *Exxon Valdez* oil spill.

For businesses that require heavy doses of startup capital, SAIC's model of employee ownership may not raise enough. But for knowledge-driven companies looking for new ways to motivate staff, tying stock to performance and spreading the wealth as widely as possible has a compelling logic. Robert Vickery, cofounder of Sparta Inc., an engineering services company in Orange County with a compensation system modeled partly on SAIC's, sums up why employee ownership works: "Nobody washes and waxes a rental car."

– Nancy J. Perry

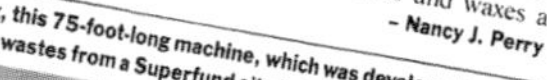

**Much like an oil rig, this 75-foot-long machine, which was developed by SAIC, pumps toxic wastes from a Superfund site in San Pedro, California.**

Reprinted through the courtesy of the Editors of FORTUNE

# WORKING FOR THE CUSTOMER

*The 1980s started with military strength as a top U.S. priority. During the dark days of the Cold War, President Reagan's commitment to a major military buildup was accompanied by the funding to support and build those programs. This sudden upturn in defense work represented a significant change in size and complexity compared to contract efforts previously required by the governmental defense agencies.*

SAIC's scientists and engineers demonstrated a high degree of flexibility by continuing the Company's earlier nuclear weapons effects work while shifting into the newer military programs calling for high levels of technical analysis, design, and integration. The list of defense research needs grew longer and included electronic warfare, infrared detection, ocean surveillance, aerospace sciences, underwater acoustics, and antisubmarine warfare systems. Support services expanded to include procedure development for laser-guided weapons and the identification of military aircraft as friend, foe, or neutral in combat situations.

In 1983, President Reagan announced the Strategic Defense Initiative (SDI), which soon became known as "Star Wars." Not only did this program call for a fundamentally new approach to defense technology, but in Reagan's view, SDI would have a monumental impact on mankind as a whole.

SAIC was chosen early in the process to play a substantial role in the research and development required for the SDI. In addition to research and development, much of the work involved designing a new system with inherent technological challenges unlike anything in U.S. military history. This design work required development and integration of complex technologies to determine if an array of new and different defensive weapons could effectively protect the United States against missile attack. This was fundamental architectural work to develop complex new systems, heavily grounded in solid science and engineering. The research and development expertise and skills of SAIC employees

SAIC develops advanced command, control, and communications systems. These systems include sensor systems, satellite systems, and communication systems, below right, that link field commanders on ships, planes, and submarines. In 1984, SAI was merged into Science Applications International Corporation (SAIC), below left, reflecting the expansion of the Company's business beyond national boundaries.

SAIC divers help prepare a submarine to run underwater noise trials on the SAIC-installed range at the Santa Cruz Acoustic Range Facility off the California coast.

were crucial factors in enabling SAIC to work successfully as a system architect for the overall SDI program.

As the SDI program evolved, SAIC continued its wide-ranging technological support for nearly all aspects of the program. The Company took a major step in evaluating the computers and software that would control the entire SDI system by building the Strategic Defense Simulation Laboratory, an advanced facility for simulating SDI battles. Both the Army and the Air Force contracted with SAIC to design and evaluate systems' survivability, assuring their ability to operate in space in a nuclear environment.

Because of the varied defense work SAIC had done prior to SDI, the Company was able to mobilize scientists and engineers to work on a wide range of concerns for the SDI program: high-energy lasers, analysis of missile threats, assessment of potential Soviet responses, evaluation of the vulnerability and survivability of space vehicles, and numerous command, control, and communication networks. The work combined analysis, technology, and planning, and disparate parts of the Company were brought together to form the teams to work on the project.

As the new military technologies of the 1980s continued to develop, SAIC's business grew at a rapid rate. By 1985, revenues had surpassed $400 million with national security contracts accounting for two-thirds of that revenue. And while the Company was moving dramatically into new and often uncharted areas of technology and architecture for national security systems, it was continuing to build on strengths that had been its backbone since its earliest days.

SAIC simulations, below, supported a wide range of technologies and systems. Using a tactical simulator for the Patriot Air Defense System, left, the Company assisted the U.S. Army in analyzing new software approaches for defeating tactical ballistic missiles.

# War Games with Lasers and Videotape

In the late 1970s, the U.S. Army built its National Training Center (NTC) at Fort Irwin, California, and began what, at the time, was the most advanced high-technology training effort in the world. From the start, the intention was to use this remote desert location to provide the most realistic battlefield experience that could be simulated outside the actual combat arena. Soldiers, leaders, and units would get a realistic test to hone their combat skills.

SAIC's role was to integrate operations, hardware, and software to create the most advanced interactive system ever developed for army training. The system's components ranged from sophisticated electronic monitoring and control systems to display consoles that could be operated by people with no knowledge of computers. Replays of battle action and real-time casualty assessments provided maximum military training benefits. Learning how to fight and survive before facing live ammunition gave the soldier a realistic experience and a better chance at winning in actual combat.

NTC allowed the army to train for the first time as it expected to fight: intense, fast-paced action requiring quick assessments and split-second decisions. Previously, this had never been possible outside actual combat situations. SAIC's systems succeeded in performing a very complicated job—tracking all of the action on the ground during simulated combat.

Throughout the 1980s, SAIC updated the various component systems of the NTC. When allied forces launched their attack against Iraq in 1991, most of the U.S. Army units that participated in the ground war had trained at the NTC. They were ready. Their training with laser weapons and videotaped battles contributed to the success of the combat effort.

For the U.S. Army's National Training Center, SAIC integrated state-of-the-art hardware and software to produce the most advanced interactive system ever devised for Army unit training.

As the U.S. Navy's largest metrology contractor in the '80s, SAIC maintained the extensive test equipment needed to keep ships, weapon systems, and aircraft, such as this F-14, operating at peak performance.

# Ethics and Quality: Good Values, Good Business

From the inception of the Company, employee ownership and high-quality work were two key components of SAIC's success in building its business. SAIC's founders were more interested in creating a comfortable and constructive working environment conducive to performing superior technical work than maximizing financial returns. Thus, the long-term impact of day-to-day actions was critical, and ethical behavior a necessity. How results were achieved was as important as the results themselves in sustaining the Company's business.

In the early years, the number of business units comprising SAIC was small and the number of employees was low enough for them to be closely bound together by common business goals. By the mid 1980s, however, the Company had over 5,000 employees in offices throughout the United States and other parts of the world, and nurturing the Company's core values had become more complicated.

In 1985, the Department of Defense launched the Defense Industry Initiative to establish an ethics program for defense contractors. Although ethical behavior was always an important part of the SAIC culture, the Company subscribed to this initiative and redoubled its efforts. SAIC's Standards of Business Ethics & Conduct were put in writing for all employees. SAIC began its ethics program as a corporate function, but the program quickly gained the support of line managers who ran individual business units. These managers realized the business necessity of adherence to a structured set of requirements. Over time, the ethics program became a partnership between a corporate committee and the line managers.

The standards are disseminated each year to each employee and cover topics ranging from time records and conflicts of interest to integrity in scientific research. Employees are required to disclose any misconduct they observe, and there are various ways to report and resolve problems. Talking to a supervisor or a local human resources manager is encouraged, but employees can also call the Ethics Committee, call an Ethics Hotline, or write a letter to the CEO.

Employees are required to read the standards each year before their salary reviews, and each employee signs a statement agreeing to comply with the standards. By the late 1990s, the SAIC team of employees totaled more than 38,000 and the ethics program had become a well-established part of the Company's operations. Twenty-five people from many different SAIC locations serve on the Ethics Committee, training employees and supervisors and investigating reports of misconduct. While SAIC employees have the right to report misconduct anonymously, over half of the reports are made with full disclosure of the person making the report. Ethics is serious business at SAIC.

SAIC's computerized Composite Health Care System, right, puts patient information at the fingertips of hospital staff. This automated medical information system was installed in hundreds of military hospitals and clinics around the world.

## Health Care Information Leaps Forward

By the early 1980s, SAIC focused much of its health care work on developing systems to improve the quality, usefulness, and accessibility of medical data. Health care costs were a concern in all segments of the U.S. economy, and the U.S. Government was eager to find ways to shift from largely manual, paper-based information systems to computerized systems. Military medical facilities used cumbersome, expensive, labor-intensive systems in which instructions were passed along on paper records.

During 1986, the Department of Defense (DoD) awarded four different companies competing contracts for the first phase of Composite Health Care System (CHCS). The DoD planned to downselect to two competing companies for test and evaluation, and then to one company for full-scale development, although it had the flexibility to select one company if its work was clearly superior. SAIC was one of four winners in the first step in what would become a huge, long-term program that ultimately encompassed both Government and private-sector health care customers. The CHCS Phase I contract involved implementing a computerized medical information system. While SAIC had extensive experience with other types of information systems, it had never installed a medical information system in a hospital or clinic.

While SAIC was integrating its system at Fort Knox, three other contractors were installing competing medical information systems at three other Government facilities as part of a Government-sponsored competition. The competition reward was a huge Government contract for a computerized medical records system for more than 500 military medical facilities worldwide. SAIC moved quickly and aggressively to create a team of experts to install a system built on the foundation of an existing Veterans Administration system. That proved to be a key advantage in the competition for the larger contract. SAIC was still the dark horse contender facing other contractors who had more extensive experience in medical data systems. But none of the competitors had honed their systems into a smoothly functioning program as SAIC had done. SAIC's approach allowed the DoD to leverage years of previous government-funded development work, save significant costs, and start with proven technology.

When the DoD prepared to award the contract for the new CHCS, it had to be sure the chosen system would meet thousands of technical specifications and accommodate users ranging from large facilities such as Walter Reed Army Hospital to small clinics in rural areas.

Neurosurgeons can now see the location of their surgical instruments superimposed on an MRI image of a patients brain, thanks to a system developed with SAIC expertise. Opposite page, SAIC health expertise and technical innovation combine in systems that support the clinical environment and improve care for millions of patients.

SAIC undertook a massive simulation to show that its solution would work technically and be less expensive.

SAIC was judged to be clearly superior, thus the DoD awarded the complete contract to SAIC in 1988. The United States General Accounting Office (GAO) reviewed the award decision and determined that SAIC "received a higher quality assessment in three of the four areas, while proposing a lower cost alternative."

The GAO monitored SAIC's progress and results during the entire development and implementation of the system, as well as during later system upgrades. In 1996, the agency reported, "GAO has been monitoring and reporting on CHCS since August 1985 . . . As the backbone of Defense's medical operations, CHCS will provide personnel with almost instant access to patient information, from medical history to current treatment and vital statistics."

The billion-dollar contract for the implementation of the Composite Health Care System for the Department of Defense came early in 1988 and was the largest in Company history. It was a milestone in the Company's history, both because this contract was greater than the total revenues of the Company, and because this contract, together with the Saudi Navy contract, put SAIC on the map as a systems integrator. In both cases, through strong performance, SAIC continued to provide these customers with systems improvements and operations and maintenance support through 1999. Along with other systems integration projects, the Composite Health Care System laid an important foundation for future work in the emerging world of information technology.

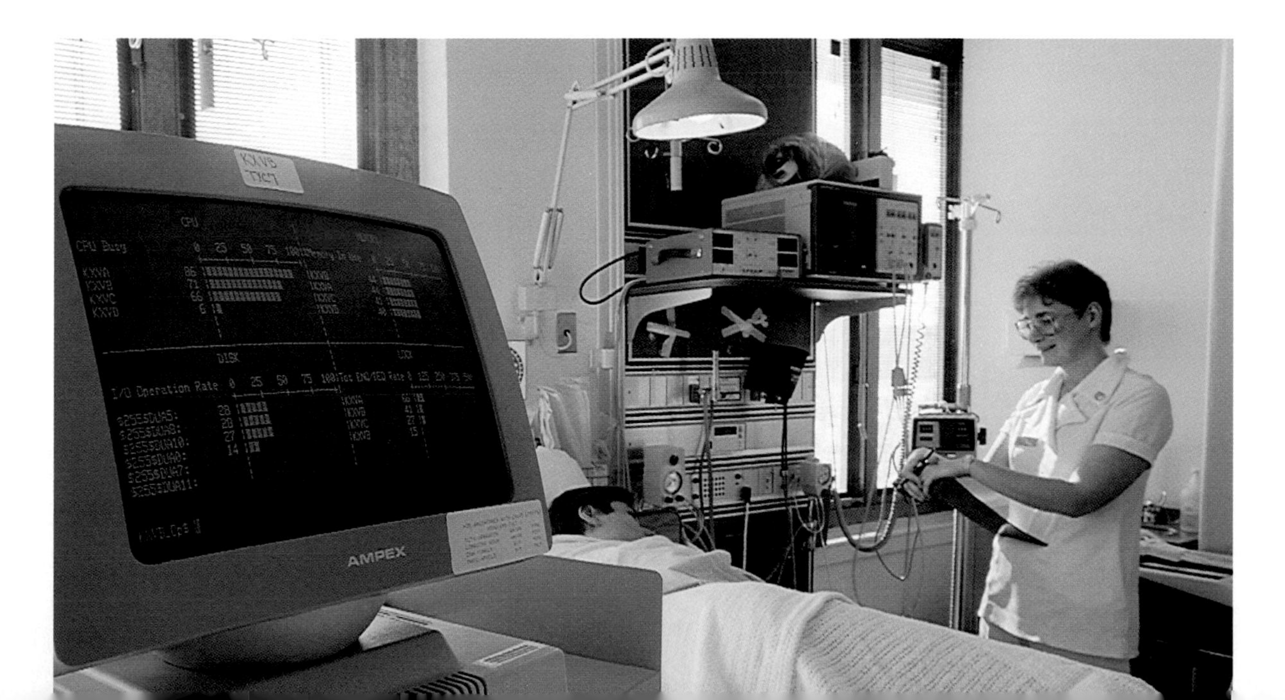


**PRINCIPLES & PRACTICES**

**Taking a long-term view of what is important for SAI, the quality of our technical efforts must always come first. Not only those efforts on behalf of our customers, but also those using our own resources for internal research, in managing ourselves, and in performing all of our administrative and business functions.**

*From "Principles & Practices of SAI," written by J.R. Beyster in 1983 to codify the Company's core values in a pamphlet for all employees.*

# Employees Gain a Formal Voice: The Technical Environment Committee

By 1980, with the Company's significant growth, employees were looking for more efficient and effective ways to share their ideas with management relating to the work environment.

One approach was to have shareholders vote a non-management employee onto the Board. Bob Beyster questioned the effectiveness of having a single Board member from the non-management ranks. Instead, he proposed to form the Technical Environment Committee (TEC) with a mission of helping SAIC foster an excellent working environment. He suggested that its chairperson attend SAIC's Board of Directors meetings. The TEC was born in 1982 in a way that provided employees with another way to express their ideas and insights.

The TEC has operated as a cross section of SAIC's non-management employees. The Committee Chair attends Board of Directors meetings. The TEC serves as a crucial conduit of communication both ways—from the Board and senior management to the employees, and from employees back to management. But rather than simply facilitating the flow of information, the TEC also brainstorms with management to build a better working environment. Beyster has called the TEC "the conscience of SAIC."

A good example of the TEC's success was seen in the early 1990s in the changes which were made to the health insurance plan.

The TEC's input provided management with additional information to evaluate SAIC's health insurance program. Based upon all the information available, management decided to implement a self-insured program, accompanied by a major educational program to inform employees that money spent on their health insurance affected the Company's results and their wealth as employee-owners. Insurance reimbursements declined, there was no need to increase the overhead costs passed on to customers, and employees used the TEC to learn about and impact the structure of that system.

TEC
Technical Environment Committee

## The Thaw

By the mid-1980s, large-scale efforts were underway to negotiate with adversaries in an effort to slow the buildup of expensive and powerful nuclear weapons. For several years, SAIC had been involved in major research regarding Soviet nuclear weapons and military operations, and the Company helped U.S. policy makers understand the implications and effectiveness of Soviet systems and technology. Soviet systems, procedures, targeting techniques, and reserve nuclear forces were topics that needed to be well understood as negotiations led up to the Intermediate-Range Nuclear Forces Treaty signed by President Ronald Reagan and Mikhail Gorbachev in 1987. Eliminating the deployment of an entire category of missiles, the treaty created a new need for arms control compliance assessments and for ongoing analysis of the treaty's impact on U.S. defense capabilities.

SAIC had learned to adapt itself to the marketplace. The Company moved quickly into the new world of architecture and design for SDI, maintained its long-standing involvement in the

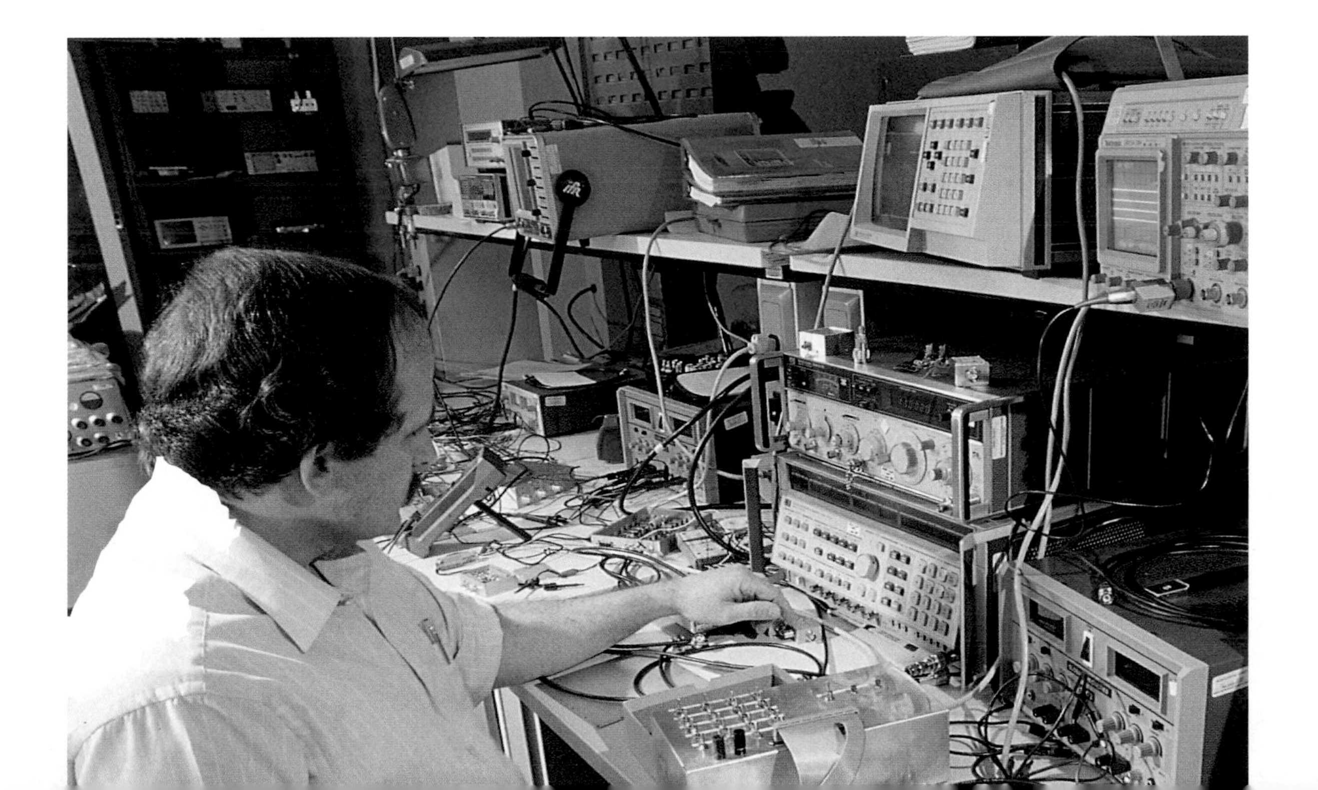

In 1986, B-1B simulators developed by SAIC began training flight crews at Dyess Air Force Base and setting new standards for performance, opposite. The following year, U.S. President Ronald Reagan and Soviet General Secretary Mikhail Gorbachev signed the treaty to eliminate intermediate-range nuclear missiles, top right. Meanwhile, SAIC's electronic combat laboratory supplied the intelligence community with prototype electronic systems, left.

SAIC's portable, combat-rugged communications work stations, below left, have served the needs of all U.S. military services. In 1986, the same year these portable computers were introduced, bottom left, SAIC Sovietologists analyzed Soviet military strategy at the Company's Foreign Systems Research Center, right.

research and development world of nuclear weapons' effects technology, and broadened its sophisticated policy analysis to include new regions such as Asia and new areas such as weapons proliferation.

As the nature of the work changed, the way the work was done also changed. Prior to the Competition and Contracting Act of 1985, government agencies customarily contracted for research and development work on a sole-source basis. Sole sourcing essentially meant that the government could negotiate with a single contractor when justified, and contractors could propose relevant, but unsolicited, work directly to government agencies.

SAIC technicians assembled highly reliable microcircuits to replace obsolescent semiconductors.

At a pulse laser laboratory, SAIC scientists developed a device to protect optical sensors from hostile laser threats.

SAIC's technical systems and products have been originated to meet specific customer needs. The Company's portable, combat-rugged communications work stations, below left, have served the needs of all U.S. military services.

When the Competition and Contracting Act replaced the sole-sourcing arrangement with competitive bidding processes, SAIC went through a major learning experience. Competitive bidding shifted more of the financial risk for first-of-a-kind development projects to the contractors. SAIC focused an internal team on developing a formalized process to help SAIC employees learn the new skills and management processes required for competitive bidding and fixed-price development contracts.

As one of the leading companies in the U.S. for automatic target recognition algorithms and software, SAIC works with various kinds of imaging sensor data.

For more than two decades, SAIC scientists have explored and developed new technologies to assist NASA. SAIC provides key technical support to two of NASA's most important efforts—operating the Space Shuttle and building the International Space Station. SAIC is the integration contractor for the Air Force's Space and Missle System Center's advanced programs, and helps the Air Force Space Command plan, operate and support their military space systems.

# Taking Care of the Neighborhood

The ban of oil exports from the Middle East to the United States in 1973 produced long lines at U.S. gasoline stations and quickly resulted in both energy and environmental projects for SAIC. Both domestic oil production and nuclear energy generation—intended as mitigants to the energy shortage—led to important environmental contracts designed to protect the natural environment and to safeguard against catastrophes.

SAIC became actively involved in environmental impact studies in the mid-1970s that extended the Company's technical expertise. Much of this environmental analysis benefited from the Company's expertise in mathematical modeling and transferred that capability to a new group of customers. Decisions about plant site selection and pollution control systems for air, water, and noise were based on environmental impact studies, and SAIC performed this work for companies in various industries and for Government agencies.

When energy producers contemplated drilling for oil off the shore of southern California, SAIC began an extensive environmental study of the continental outer shelf. This project established the base line analysis of the coastal conditions to assess the impact of oil drilling. An SAIC-led team gathered and evaluated water samples, ocean floor samples, and various biological and chemical materials to understand and measure the status of the ocean shelf. That project continued for several years and led to further work monitoring environmental effects along the Pacific and Atlantic coasts and the Gulf of Mexico.

SAIC became involved in the nuclear power generation business in the early 1970s. The expertise developed in national security work shielding nuclear reactors was transferred to the nuclear power industry and helped make those nuclear power plants safer. Assessment of the safety and reliability of nuclear power plants was a business in which SAIC excelled. The Company also had experience with the thermal impact of waste heat being disposed in rivers, another operational concern at nuclear plants. This knowledge helped determine how to minimize the impact of heat effluents, and strengthened SAIC's reputation regarding safety and environmental issues facing the nuclear power industry.

When the Three Mile Island nuclear plant in Pennsylvania was damaged in 1979, SAIC was asked by industry and government officials to monitor radioactive releases from the plant. The Company continued monitoring work at the plant long after it had been stabilized, and that project led to other nuclear monitoring work with industry and with the Nuclear Regulatory Agency during the 1980s. Following the rupture of the Chernobyl reactor core in 1986, SAIC was asked to assess the safety of nuclear reactor facilities in the U.S.

In 1989, following the Exxon Valdez oil spill off the Alaska coast, once again SAIC scientists were called in to help determine the impact of the oil on the environment, though this time it was a subarctic location rather than southern California.

In 1988, SAIC conducted a search of the sea floor off Peggy's Cove, Nova Scotia, to ensure that all of the wreckage of Swiss Air 111 had been recovered. The search equipment featured SAIC's Laser Line Scan System, also used by SAIC in 1996 in the TWA Flight 800 crash investigation. The Laser Line Scan System, developed and used for undersea surveys, also turned out to excel at accurately locating, rapidly assessing, mapping, and providing a detailed evaluation of wreckage on the sea floor.

During the 1980s and 1990s, SAIC won contracts to provide critical technical and management support to a proposed underground repository for high-level nuclear waste in Nevada.

SAIC helped nuclear power plants around the United States increase plant availability, reduce costs, and improve safety.

**PRINCIPLES & PRACTICES**

**A special challenge to an organization as far flung as SAI is to achieve the levels of coordination and cooperation needed for efficiency. As SAI grows and the groups inevitably become more self-reliant, this challenge could become harder to meet. The Management Council, National Security Policy Group, regular gatherings of technical specialists, group meetings and intergroup support on projects all provide for a healthy interchange of information.**

*From "Principles & Practices of SAI," written by J.R. Beyster in 1983 to codify the Company's core values in a pamphlet for all employees.*

## Shifting Strategies

During the 1970s and 1980s while sole sourcing was underway, much of the Company's strategy was focused on shorter term goals—selling employees' time and marketing what could be delivered now. Over time, a more structured strategy would focus on types of business and categories of customers that were likely to produce sustainable future business, though the hustling marketing efforts of individual managers would always remain a critical factor in SAIC's growth and profitability. The desire to diversify the base of business became increasingly important during the late 1980s. Government and defense spending leveled off or declined for many of the programs in which SAIC had been deeply involved, and aggressive marketing in new areas was critical to sustain the Company's revenue growth.

SAIC managers bid competitively, and their strategy paid off. Revenues and profits grew every year, and by 1989 the revenues of the 20-year-old company stood at $865 million. Most of the contract work was the result of SAIC employees' initiative in pursuing business they felt they could win and perform. Increasingly SAIC managers worked with their SAIC colleagues in putting together more complex proposals that required interdisciplinary resources. Informal networks developed among managers as they learned whom they could rely on and how to work together. They learned to write and present good proposals, hire good people, bid often and hard, and hope for occasional luck.

In the late 1980s, SAIC increased its revenues and profits both from internal growth and acquisitions. During this decade, the largest acquisition was American Systems Engineering Corporation, known as AMSEC. This acquisition strengthened SAIC's naval logistics support business and leapfrogged the Company into a stronger and more competitive position. Based in Norfolk, Virginia, AMSEC's 280-person staff (which grew to 1,600 by 1999) performed engineering and maintenance for ships—a business that was consistent with SAIC's engineering tradition, but for a different category of customers. The AMSEC acquisition was important both in terms of the business that it generated and as a demonstration that SAIC could make significant acquisitions. The Company's acquisition strategy would become increasingly important in the next decade. In 1999, AMSEC formed a business partnership with Newport News, creating an integrated engineering, planning, and logistics entity positioned to support all aircraft carrier, submarine, and surface ship platforms.

In 1985, SAIC won a contract to build a turnkey facility for the Saudi Navy. The project, which began in 1979 as a feasibility study, was massive, requiring complex systems integration to bring together an array of systems to manage virtually every aspect of the Saudi Navy's $C^4I$ system

Providing high-quality engineering and maintenance work has helped SAIC's AMSEC subsidiary win all major recompetes without fail throughout its history.

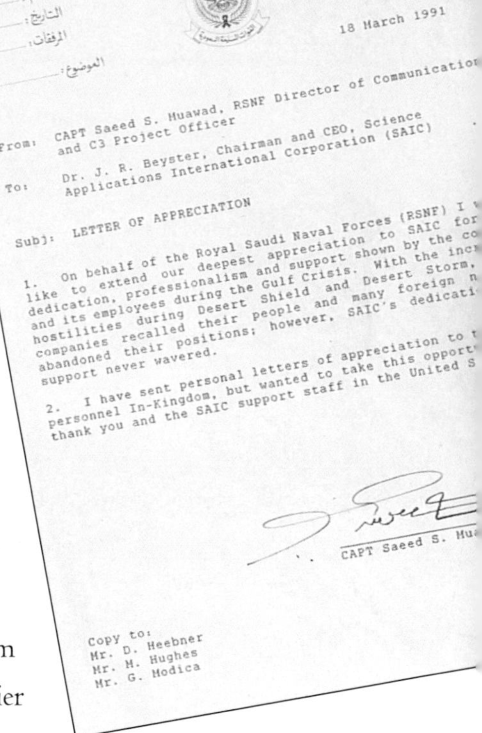

18 March 1991

From: CAPT Saeed S. Huawad, RSNF Director of Communicatio
and C3 Project Officer

To: Dr. J. R. Beyster, Chairman and CEO, Science
Applications International Corporation (SAIC)

Subj: LETTER OF APPRECIATION

1. On behalf of the Royal Saudi Naval Forces (RSNF) I
like to extend our deepest appreciation to SAIC for
dedication, professionalism and support shown by the co
and its employees during the Gulf Crisis. With the inc
hostilities during Desert Shield and Desert Storm,
companies recalled their people and many foreign n
abandoned their positions; however, SAIC's dedicati
support never wavered.

2. I have sent personal letters of appreciation to t
personnel In-Kingdom, but wanted to take this opport
thank you and the SAIC support staff in the United S

CAPT Saeed S. Hu

Copy to:
Mr. D. Heebner
Mr. H. Hughes
Mr. G. Modica

SAIC personnel in Saudi Arabia earned lavish praise from the Saudi authorities after supporting the operation of one of the most important command and control systems in the country, even after Riyadh came under fire. The development and integration of that system, far right, was one of the Company's top achievements in the 1980s.

operations. Support to this customer would result in almost $400 million in revenues for SAIC thoughout the 1980s.

SAIC assembled a team of specialists from various parts of the Company, a team that continued to expand as the full scope of the systems integration requirements became apparent. The extensive resources of SAIC were crucial in designing and implementing this facility, and the work was highly successful. Years later during the Gulf War, this command, control, and communications center became the focal point for Saudi Arabia's national defense efforts. SAIC employees continued to assist in operating the center throughout the 1991 hostilities, drawing praise from Saudi authorities.

This systems integration work was fundamentally different from most of the other projects SAIC had done. While much of SAIC's earlier work involved research focused on complex issues, the Company's systems integration work involved assembling hardware, software, and operating systems in innovative ways. It also involved greater risk because usually more money was spent up front putting together the design approach and the proposal and negotiating with other contractors and vendors.

SAIC managers saw that this type of work was becoming more prevalent among both government and private-sector customers, and when done well, was profitable. While the systems integration project for the Saudi Navy moved SAIC to the forefront of this type of business, it also built on the science and engineering work, the research and development work, and the design work for which the Company was known.

Though many details of the Saudi work could not be revealed for security reasons, the success of the facility was known and potential customers saw SAIC as a serious player in the field of systems integration. Much of the Company's work during the balance of the 1980s involved the integration of multiple systems for various clients; the Saudi contract proved internally and externally that the Company could excel in this field. SAIC's support to the Saudi Navy was ongoing as of 1999—one example of SAIC's long-term commitment to its customers.

CENTER STATUS
SAIC

After 19 years of renting space in numerous buildings in La Jolla, the Company moved its corporate headquarters to Campus Point in San Diego, constructing seven buildings there. This was SAIC's first foray into building ownership.

When *Stars and Stripes* won back the America's Cup from Australia in 1987, it was more than a sailing victory. It was also a technology victory for SAIC. The highly publicized win capped three years of computer hydrodynamic simulations, design tradeoffs, and testing by SAIC. The Company's efforts pushed forward the state of the art in yacht design and helped produce a technically superior boat.

SAIC™

# *Stars and Stripes*: Victory at Sea and in the Research Lab

When the 12-meter yacht *Stars and Stripes* battled to win back the America's Cup from Australia in 1987, it was a technological victory for SAIC as well as a national victory in the world of sailboat racing. SAIC scientists and engineers worked for three years on the intricacies of yacht design and performance evaluation. Before the yachts ultimately did battle at sea, some of the best design and technical engineering minds around the world competed on the drawing boards to create a technologically superior boat.

SAIC had analyzed ship performance over many years for the U.S. Navy, but the Company had never undertaken the design and performance evaluation for 12-meter yachts. What SAIC did have was the determination to leverage its existing knowledge and skills to meet this new challenge. From computer simulations in the research lab to the sailing trials off Hawaii, this team of scientists and engineers built on what they knew and added new knowledge gained from first-hand experiments.

As the key technical support contractor for *Stars and Stripes*, SAIC contributed to major breakthroughs in hull and keel design and in the performance evaluation of this class of boat. The design work for *Stars and Stripes* put the SAIC team in the most visible public arena in which they had worked: millions of people around the world would know whether they won or lost. In 1999, SAIC was providing technological support to the San Francisco-based syndicate, America One, hoping to help the U.S. recapture the Cup in 2000 after the loss to team New Zealand in 1995.

DR. J. ROBERT BEYSTER
IN APPRECIATION OF YOUR STEADFAST
SUPPORT IN DEFENSE OF THE AMERICA'S CUP
CROSS SECTION FROM US-1 CATAMARAN DAGGER BOARD
SAN DIEGO 1988

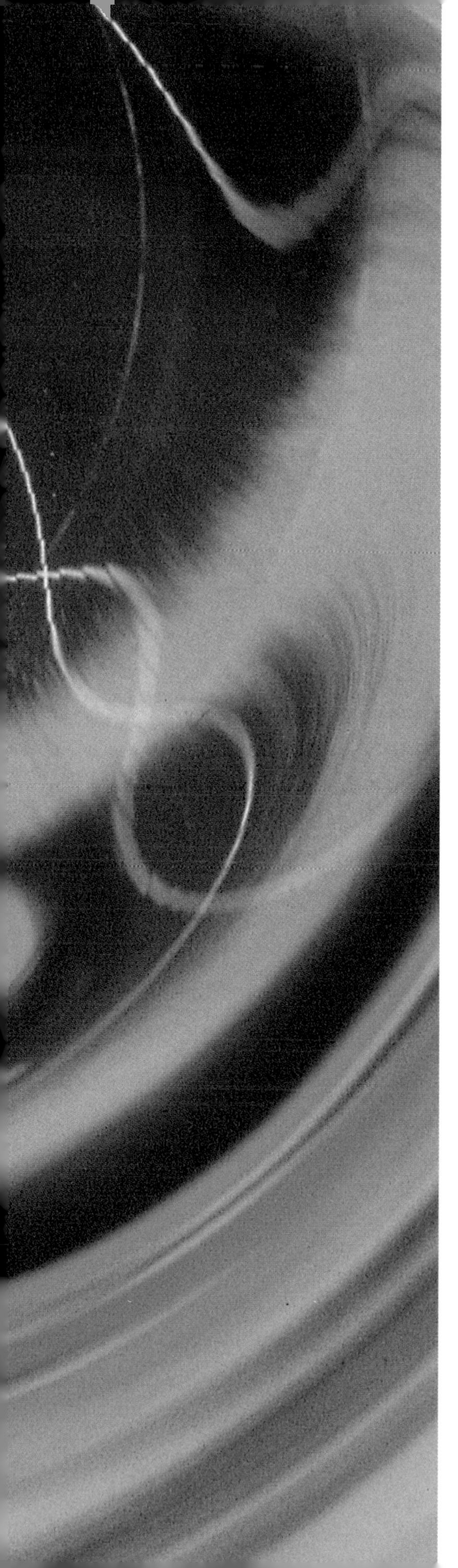

# Pushing into New Markets

*In 1993, SAIC's commercial and international non-Government business was about 10 percent of the Company's total revenues of $1.5 billion. In spite of a leveling or decline in Government spending, SAIC gained an increasingly larger share of Government business, fueling the Company's continued growth throughout the 1980s. But during the early 1990s, the senior managers at SAIC saw persuasive reasons to diversify into more non-Government business.*

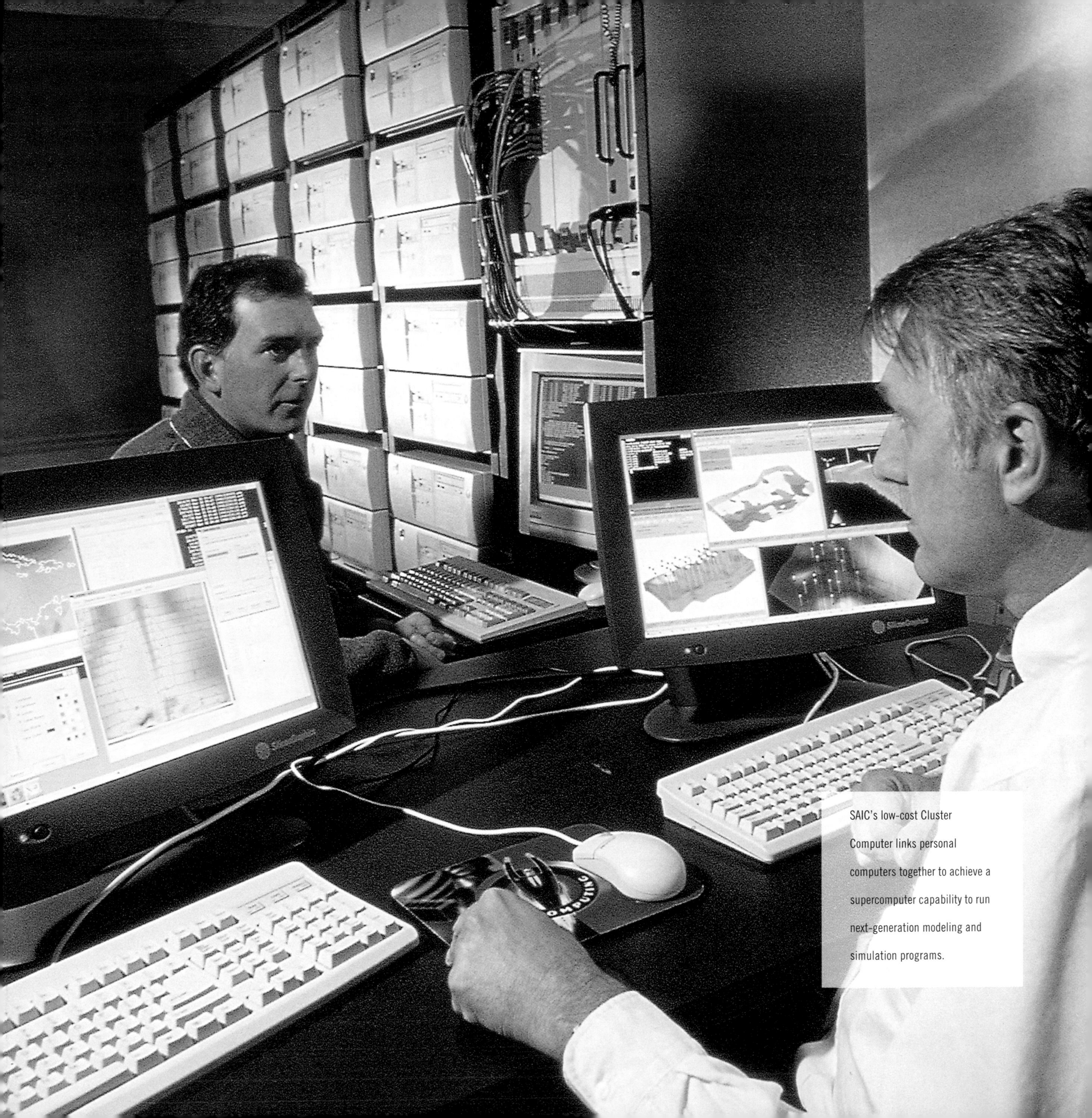

SAIC's low-cost Cluster Computer links personal computers together to achieve a supercomputer capability to run next-generation modeling and simulation programs.

First, while the Company's market share of national security spending increased steadily, there were concerns that over the long term, federal budget reductions would inevitably slow SAIC's growth. Second, Company employees had an extensive base of technical skills that were as applicable for commercial customers as for Government customers. Third, many SAIC managers were eager to undertake new challenges with commercial work.

The initial commercial strategy was to focus on information technology (IT) because it was growing at breakneck speed in the marketplace, and because that business built on SAIC's technical and managerial strengths. IT involved solving complex technical problems with sophisticated and often innovative solutions, and difficult problem-solving had historically been the core of SAIC's business. In addition, SAIC had already established a strong reputation in one of the key components of IT—systems integration—through its work on large and successful Government programs. Starting in the early 1990s, SAIC set its sights on becoming an IT leader in the commercial arena as well. SAIC's leaders felt the Company had an edge over other IT companies because SAIC had great depth of technical skills, and the Company believed its employee-owners worked harder to satisfy customers.

SAIC selected the oil and gas industry as an initial target for IT business because many of the leading companies in the industry were going through significant strategic changes and were looking to information technology to improve their productivity and decrease their costs.

### In the United Kingdom

In 1992, British Petroleum decided to outsource its information technology operations and introduce new forms of technology. Outsourcing at that time meant having another company provide technology services under a long-term, fixed-price contract and actually manage the computers, the information systems, the people, and that part of the company's operations. Outsourcing involved a large amount of trust between the customer and the contractor since the customer relied heavily on crucial computer operations to focus on its own business strategies.

When British Petroleum first sought bids for its IT outsourcing contract, 67 companies were considered. The group was narrowed to 15, to 6, and finally to 3, and throughout the process SAIC was a dark horse contender because it had not done IT outsourcing work on this scale before. But SAIC's strength was its familiarity with technology in fields other than oil and gas—British Petroleum was seeking new ideas for using technology—and its willingness to undertake a new way of structuring the outsourcing contract.

**PRINCIPLES & PRACTICES**

**Growth is necessary for maintaining the right working environment in SAI. It provides opportunities for individuals to expand their technical areas of interest and to advance in management responsibility. Growth, of course, is also important in creating the financial rewards necessary to attract and hold the best people.**

*From "Principles & Practices of SAI," written by J.R. Beyster in 1983 to codify the Company's core values in a pamphlet for all employees.*

YANBU
WARNING
115 VOLTS
ABQAIQ

SAIC helps protect information assets for oil companies and power plants. This system controls access for a quarter of a million Arabian-American Oil Company (ARAMCO) employees.

From reengineering business processes to developing computing infrastructures, the Company's INTESA joint venture in Latin America, right, assists industries in finding innovative ways to lower costs.

Rather than stay with the traditional long-term, fixed-price arrangement then common for commercial IT customers, SAIC agreed to a risk-reward sharing model. This concept did not have a fixed price, but instead involved setting financial targets each year and sharing money saved or spent compared to that target with the customer.

The willingness to share risks and rewards and the compatible cultures of British Petroleum and SAIC moved the Company from being a long-shot bidder to being one of the winners. SAIC became one of three companies selected as worldwide partners in operating British Petroleum's computer facilities and operations. The Company began its work with British Petroleum managing the oil company's computer operations in Aberdeen, Scotland.

With this contract, SAIC began a substantial, ongoing relationship with British Petroleum and established itself in the top ranks of IT providers to commercial customers. Once again the Company built on its technical and cultural strengths to bootstrap itself into a new business.

**In Venezuela**

In the mid-1990s, one of SAIC's petroleum industry experts went to Latin America in connection with work the Company was doing for British Petroleum. During his work there, the SAIC manager got to know senior management at Petróleos de Venezuela, S.A. (PDVSA), a national oil company and the world's second-largest oil producer. PDVSA sought bids from contractors in 1996 to provide information technology, and the competition was ultimately narrowed to IBM and SAIC.

PDVSA selected SAIC, and the two companies crafted a joint venture model for doing the IT work. In Venezuela, SAIC and PDVSA formed a joint venture known as INTESA. The joint venture commenced its IT operations with 1,500 employees as one of the largest IT companies in Latin America.

Both SAIC's diverse technological background and its employee-ownership culture were important factors in the decision to establish INTESA as a joint venture and a separately owned company. When this new company was formed, employees of the new joint venture faced two significant changes: shifting from working for an internal IT organization of a major corporation to being part of a service-oriented, value-adding IT private company and having the opportunity to become employee owners. Their response to these changes was overwhelming—more than 80 percent of the employees bought SAIC stock on their first opportunity to do so—boldly introducing employee ownership as a way of doing business in Latin America.

PERL 5

The formation of INTESA in 1997 was a significant departure for SAIC. For the first time, the Company established a separate entity, of which it was a partial owner, to conduct all of its IT work in Latin America. INTESA would handle not only the IT work for PDVSA, but it would also do IT work for other companies in Latin America. The benefit to PDVSA was that it would continue to own a portion of INTESA, and new technology introduced by SAIC for the growing joint venture would be available to optimize operations, reduce costs, and add value to PDVSA and to other IT customers.

Throughout the 1990s, SAIC continued to build its IT business with commercial customers. At the same time, the Company expanded its commercial and international business through strategic acquisitions in the fast-growing telecommunications industry. By 1999, SAIC's overall commercial and international business had increased to 45 percent of the Company's revenues, up from 10 percent just six years earlier. While U.S. Government business continued to grow in absolute dollars, the commercial and international business was growing at a much faster rate. SAIC's diversification strategy was working, and an international foundation was being built for future growth.

Dedicated to finding better ways to gather, use, and apply knowledge, SAIC's INTESA joint venture, left, supports the information technology needs of major companies in Latin America. INTESA's data center is a hub for its business activities, extending throughout Latin America.

# Spreading the Word: the Foundation for Enterprise Development

*"To foster the development of productive, competitive enterprises by promoting entrepreneurial employee ownership and participation as a fair and effective means of motivating the workforce and improving corporate performance.*

*By providing practical information and assistance to help organizations use equity-based compensation and broad-based participation program effectively, we work to enhance economic and social development worldwide through broader ownership and involvement in the free enterprise system."*
—FED Mission Statement

The concept of employee ownership that Bob Beyster firmly instilled in SAIC in 1969 has not only served SAIC well but has also attracted attention from business leaders and scholars. Beyster established the non-profit Foundation for Enterprise Development in 1986 because he believes that companies that share ownership and empower employees to work effectively will be better equipped to compete in the world marketplace.

The Foundation's activities focus on providing practical information and assistance to help companies implement effective equity compensation and employee involvement strategies. Much of the Foundation's work involves advising business owners and managers as they develop and implement employee ownership strategies for their companies.

The Foundation's work has greatly expanded over the last several years, but its cornerstone continues to be providing complimentary consulting services to business leaders interested in implementing an employee ownership plan. In addition to the consulting, the Foundation has established itself on the internet (www.fed.org) as a leading site for information on employee ownership, equity compensation, and international privatization efforts, with over 18,000 visitors each month. They developed and market multimedia products, selling over 2,000 items in less than two years. Regular workshops and a yearly conference on leading employee ownership and management issues are conducted, attracting roughly 1,000 business leaders each year. Additionally, the Foundation hosts entrepreneurial training programs both here and abroad to guide hundreds toward a better

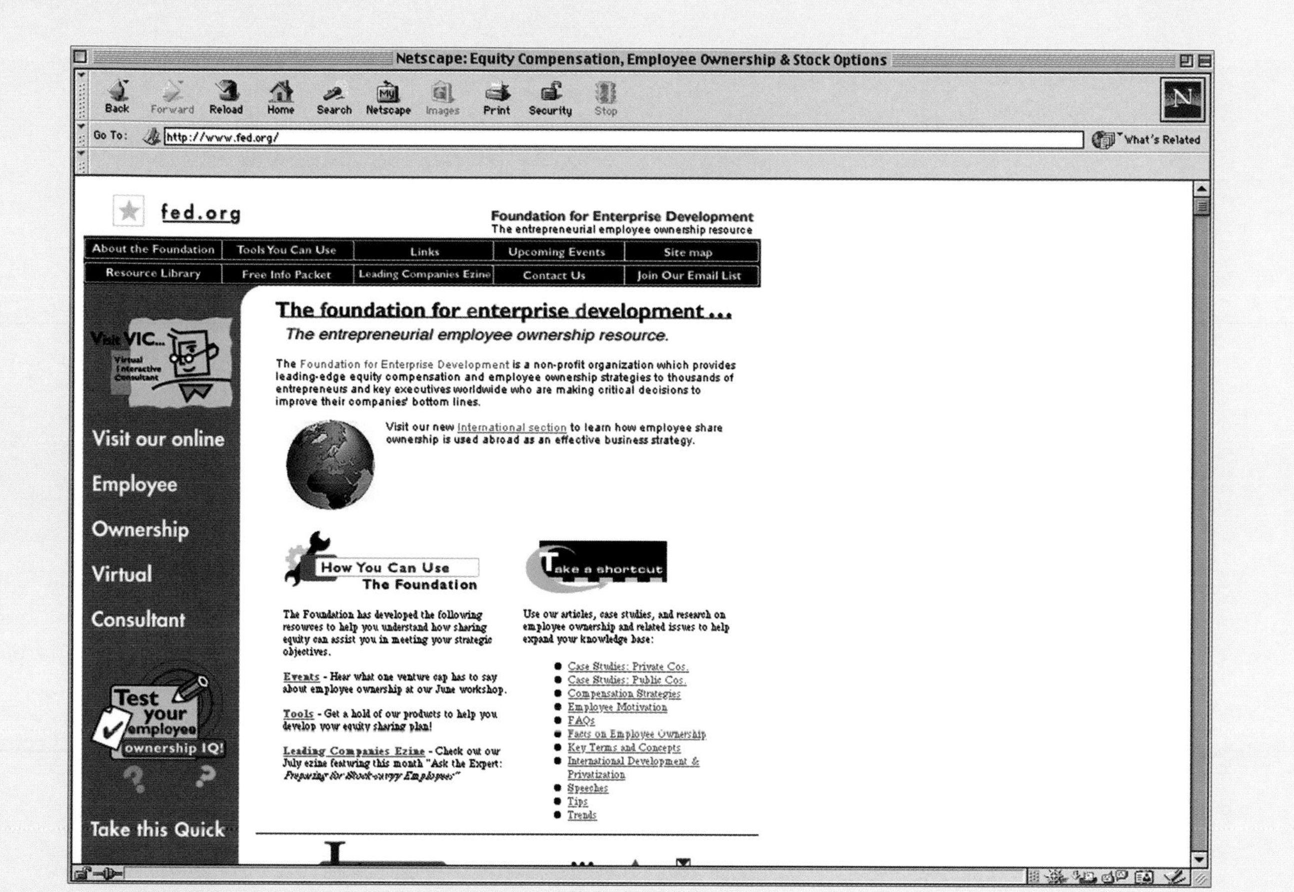


The Foundation has sold over 2,000 products, far left, to help entrepeneurs establish employee ownership programs. The Foundation's extensive resource on the World Wide Web receives over 18,000 visits per month.

understanding of effective business strategies.

The impact of the Foundation is far reaching. In addition to advising more than 150 companies a year in the United States, Foundation members work with foreign government officials on privatization transactions and travel the globe from Russia to Zimbabwe, Venezuela to Egypt to provide technical assistance on employee-ownership programs. Members instill a strong sense of the power of employee ownership, and how it can be used to strengthen an organization and facilitate economic development.

The Foundation for Enterprise Development was established on the belief that equity compensation is a critical component of successful enterprise-development strategies. As the concept of employee ownership continues to gain attention in both public and private sectors as a way to more effectively manage a wide variety of business and governmental activities, the Foundation continues to promote the vision of entrepreneurial employee ownership as a fair and effective way to promote enterprise development.

SAIC offers a full range of intranet (internal network) services to help clients reengineer their operations for the electronic business environment.

## Entering Commercial Telecommunications

In the early 1990s, SAIC's leaders felt the time had come for the Company to commit itself to becoming a major player in telecommunications, an industry in which both the technology and the players were rapidly transforming what had been a predictable business into an arena of unprecedented change. SAIC had already developed telecommunications network systems for Government customers and was in the process of integrating the largest medical telecommunications in the U.S. for the Department of Veterans Affairs. As SAIC developed its commercial telecommunications strategy, the Company saw both existing customers and prospects who needed consulting help with complex telecommunications networks. But the practical problem was resources. SAIC had no extensive experience in this business and, therefore, had few people skilled in the particulars of telecommunications consulting.

SAIC acquired Network Solutions, Inc. (NSI) in 1995 because NSI had a large number of commercial telecommunications and computer consultants based in the Washington, D.C., area. NSI's telecommunications consulting work for companies such as NationsBank and AT&T fit with SAIC's game plan, and NSI's cooperative agreement with the National Science Foundation to register Internet domain names ending with *.com, .org, .net,* and *.edu* put NSI at the heart of the newly developing world of the Internet.

SAIC immediately provided additional capital to NSI, which was not making a profit. The fresh capital was invested in building NSI's computer infrastructure to process the growing business of registering Internet names. It also provided motivation to NSI employees for whom financial stability meant a more positive future. In 1995, the National Science Foundation, which had previously paid for the registry of Internet names, modified its original cooperative agreement and required NSI to begin charging annual fees to registrants.

### On the Web

The cooperative agreement with the National Science Foundation to register Internet names was not a large part of NSI's business prior to its acquisition by SAIC because the World Wide Web was just beginning to become a factor in the communications arena. NSI had one employee who registered Internet names (usually several hundred a month), and no one could anticipate the fundamental and far-reaching changes the Internet would bring to the way people and businesses interact.

SAIC realized that substantial amounts of capital would be required to sustain the growing Internet registration business since both additional computer hardware and complex software were

**PRINCIPLES & PRACTICES**

**A hard and fast, unchanged organizational structure with a fixed-block diagram will never fit the overall needs of SAI. In keeping with this organizational concept, SAI has successfully put in place a lean and sensible set of administrative procedures and signature authorities to control a very distributed operation. Furthermore, we believe that leaving as much autonomy at the operating level, and providing as many discretionary resources as possible to those who generate them is the best approach for the long term.**

*From "Principles & Practices of SAI," written by J. R. Beyster in 1983 to codify the Company's core values in a pamphlet for all employees.*

A breakthrough by SAIC scientists could significantly increase data transfer rates in optical computing. The Company's scientists, right, helped demonstrate, for what may be the first time, how to project holographic images in liquid crystal materials and repeatedly switch those images on and off by switching an external electrical field.

required. For the first time, SAIC was entering a transaction processing business (as contrasted with its contract or consulting businesses), and the financial and operational ramifications of this new business were very different from the much less capital-intensive contract work that had been the lifeblood of the Company since its inception. SAIC had ample brain power and management expertise to nurture this business, but the huge capital requirements called for a new approach.

In 1997 NSI was profitable for the first time. That year SAIC sold 26 percent of NSI through an initial public offering, and NSI raised $52.5 million. For the first time, a part of the SAIC group of businesses had public ownership; the valuation of NSI was determined by the price of its stock in the public marketplace. Because of its growth, profitability, and the success of its initial public offering, the market valuation of NSI was much greater than it had been as a privately held business, and the overall group of SAIC employee-owners benefited by the increased valuation of this part of their company since now the entire SAIC group was now more valuable.

By 1999, over five million Internet names had been registered by NSI, and the one employee registering Internet names had long since been replaced by a striking facility of computers and people working around the clock registering an average of 20,000 names each day. SAIC had moved cautiously into a transaction business where issues of scale, competition, privacy, trademark dispute resolution, cost, and retail marketing were unlike anything it, or most companies, had previously undertaken. In February of 1999, SAIC reduced its ownership of NSI to 45 percent through a secondary offering, which was at that time the largest-ever Internet equity offering . It raised $730 million pre-tax for SAIC.

NSI was the Company's first foray into the capital markets to raise substantial amounts of financing. But more importantly, it was an extraordinary learning experience—the Company learned that it could move quickly in a new, fast-changing business and could operate successfully in the public financial markets. NSI became a remarkable illustration of the Company's persistence and determination to succeed, even when it meant doing a different kind of business in a different way. SAIC's constant search for new ways of solving problems and determination to win turned NSI into a successful learning experience that bodes well as SAIC moves into the fast-changing commercial world.

*"SAIC had ample brain power and management expertise to nurture this business, but the huge capital requirements called for a new approach."*

# On the Forefront of Cancer and AIDS Research

SAIC manages the scientific resources and services of the National Cancer Institute's Frederick Cancer Research Center. National Cancer Institute and SAIC researchers work as a team to research the cause, treatment, and cure for cancer and AIDS, below.

SAIC's win of the contract to manage the National Cancer Institute's Frederick Cancer Research and Development Center in 1994 is an excellent example of how one person, with entrepreneurial passion and persistence, can galvanize a team into action to succeed at SAIC.

Ken Sunday, SAIC senior vice president for business development, was new to the company when he was asked to explore opportunities with the National Institutes of Health (the government agency responsible for the contract). He discovered that the National Cancer Institute was planning to recompete the largest single research contract awarded by the Department of Health and Human Services. SAIC management was not convinced that the Company should bid on this contract because, while the company had for some time engaged in biomedical research, it had never managed a job of this type on this scale. Furthermore, winning this contract would require a world class principal investigator with extensive experience in cancer and AIDS research. No one at SAIC had such credentials.

Ken Sunday teamed up with SAIC's Wayne Coleman, a corporate vice president and 24-year employee, and the two became determined to bid and win this contract. They put together a team with the support of John Nelson, an SAIC group manager. In the search for a world class scientist, Ken and Wayne had input from a number of people including Dr. Jim Wyngaarden, the former director of NIH who led them to Dr. Peter Fischinger, who was exceptionally well qualified for the job. Dr. Fischinger was vice president for research at the Medical University of South Carolina and former director of the National AIDS Program Office with the Assistant Secretary for Health in Washington, D.C. He was also a former Deputy Director of the National Cancer Institute, former Director of the Frederick Cancer Research Facility, and highly regarded by key individuals within NCI at Frederick and Bethesda.

John Nelson and several other senior SAIC staff helped convince Dr. Fischinger and the Medical University of South Carolina to join the SAIC team. The team put together a winning proposal for a contract, which would generate over $100 million in annual revenue and involved performing applied cancer and AIDS research and the management and operation of a research facility with over one hundred buildings. The contract win put SAIC on the map in the biomedical field—managing many of the assets of one of the leading cancer and AIDS research centers in the world. The contract also led to other contracts with the National Institutes of Health and was an important factor in the acquisition of Pathology Associates by SAIC in 1996.

"Although we did not have the biomedical expertise in house, SAIC is a scientific company that understands the needs of people in the research community," said Sunday. "I was confident that we could provide a stimulating, nurturing environment for the folks at Frederick."

SAIC transferred the 1,300 employees from the previous contractor at NCI Frederick to its payroll and began managing the entire facility. The staff responded well to SAIC and to its entrepreneurial employee-ownership culture. In 1999, four years into the contract, SAIC had set records on the award fee for this contract, a measurement of the customer's satisfaction with the contractor's performance. In addition, very few of the NCI Frederick staff left the center.

The NCI Frederick facility makes a difference in people's lives through its vital programs: The NCI Intramural Research Support Program has a number of outstanding young scientists making highly significant contributions, which made national headlines in the area of genes and disease susceptibility; The AIDS Vaccine Development Program operated by SAIC Frederick provides support for AIDS and human retroviral research efforts as a national resource, including large-scale HIV virus production and purification and development and maintenance of reagents for vaccine studies; The Advanced Biomedical Computing Center provides state-of-the-art computing support and technology to more than 1,500 scientists of NCI, NIH, and extramural biomedical research organizations through the Internet and fiber-optic networks; Through the NCI Frederick Biopharmaceutical Development Program, researchers develop experimental therapeutics for pre-clinical and Phase I and II clinical trials in cancer and other diseases.

# Division Manager's Forum: Enhancing SAIC's Operations

To attend an SAIC business meeting and hear discussions about the performance of various organizations is baffling to the outsider, not because the organizations are called by high-technology names, but because they are called by the names of the managers who run them. The division is the basic business unit of the Company, the fundamental profit and loss center, and the heart of entrepreneurialism within SAIC. The division managers are not temporary occupants of an institutional organizational structure with a separate life of its own—these division managers create businesses that, for the most part, are pursuing the passions of these managers. The Division Manager's Forum, convened quarterly, enables division managers to express their concerns about issues that impact how they operate. SAIC's more than 400 division managers and others within the Company interested in the particular issues up for discussion are invited to attend this open meeting.

The chairman of the Forum works with an eight-member executive committee to identify operational issues that need to be addressed. Issues in the past have ranged from systems that could speed or slow the collection of amounts due from their customers to non-monetary ways of compensating employees and alternative ways of structuring career paths.

The Division Manager's Forum is a formidable group, partly because it represents the front-line managers who are largely responsible for generating the Company's profits, and partly because they engage in spirited debate about significant issues. The Division Manager's Forum is part of SAIC's extensive committee system, which includes management level, technical, marketing, and affinity groups formed to ensure that a diversity of viewpoints are heard on issues of importance to SAIC and its employee-owners. This committee structure is the primary mechanism for ensuring broad employee participation in this employee-owned company.

Hearing SAIC's organizations at all levels identified by the last name of manager in charge is puzzling for anybody not familiar with this cultural attribute, but it reflects the manager's intensely personal connection with what has become a very large company.

## The Decline of the Cold Warrior

Senator Dick Lugar once testified before a Senate committee that prior to the collapse of the Soviet Union, he had been a "cold warrior," but the decline of the U.S.S.R. had created a new and different threat. Lugar said, "The viability of their entire weapons custodial system was no longer viable. Hundreds of tons of nuclear weapons material were spread across multiple sites in Russia and other former Soviet states. Russian leaders requested our cooperation in securing and protecting Russia's nuclear arsenal and weapons-usable materials. This was the genesis of the Nunn-Lugar Cooperative Threat Reduction Program."

With the passage of this program by Congress in 1991, SAIC undertook a new type of national security work, different in many ways from that done during the Cold War confrontation of superpowers. Much Cold War research and development was aimed at creating new military systems. The Nunn-Lugar

SAIC technology significantly increases U.S. capability to verify nuclear test limitation agreements. At the Center for Seismic Studies, an SAIC system detects very small seismic events.

Act provided funding in the early 1990s for collaborative efforts by the U.S. Department of Defense and the Russian Ministry of Defense for the denuclearization of the former Soviet Union—taking apart what America's former number one adversary had built.

The U.S. Department of Defense was not staffed to handle this work, so both the planning for and the removal or safeguarding of weapons was done by contractors. SAIC developed high-level technical and administrative plans for a variety of contractors (including some Russian firms) to clean up nuclear materials. The work was complex and sensitive—disposing of rocket fuel in environmentally acceptable ways, destroying missiles, and building secure facilities to store nuclear weapons.

U.S. and Russian officials supported the Nunn-Lugar program because it resulted in threat reduction and created a decreased likelihood that terrorist groups or unpredictable governments would have access to a massive supply of weapons and destructive materials. For SAIC, this program signaled that Government work in the 1990s would be very different from the 1980s.

The Company also integrated the International Data Center for Nuclear Treaty Monitoring starting with this prototype center.

# Recognizing SAIC's Science and Technology Leaders

The Executive Science and Technology Council (ESTC) represents the breadth and depth of SAIC's technological capabilities. Its members are world leaders in the application of science and technology—they are innovative scientists adept at envisioning how to apply new discoveries to solve important problems and improve people's lives.

The ESTC was formed in 1982 to promote high-quality, imaginative technical work and reward distinguished technical accomplishment within SAIC. While much of the Company's growth was being led by those who had

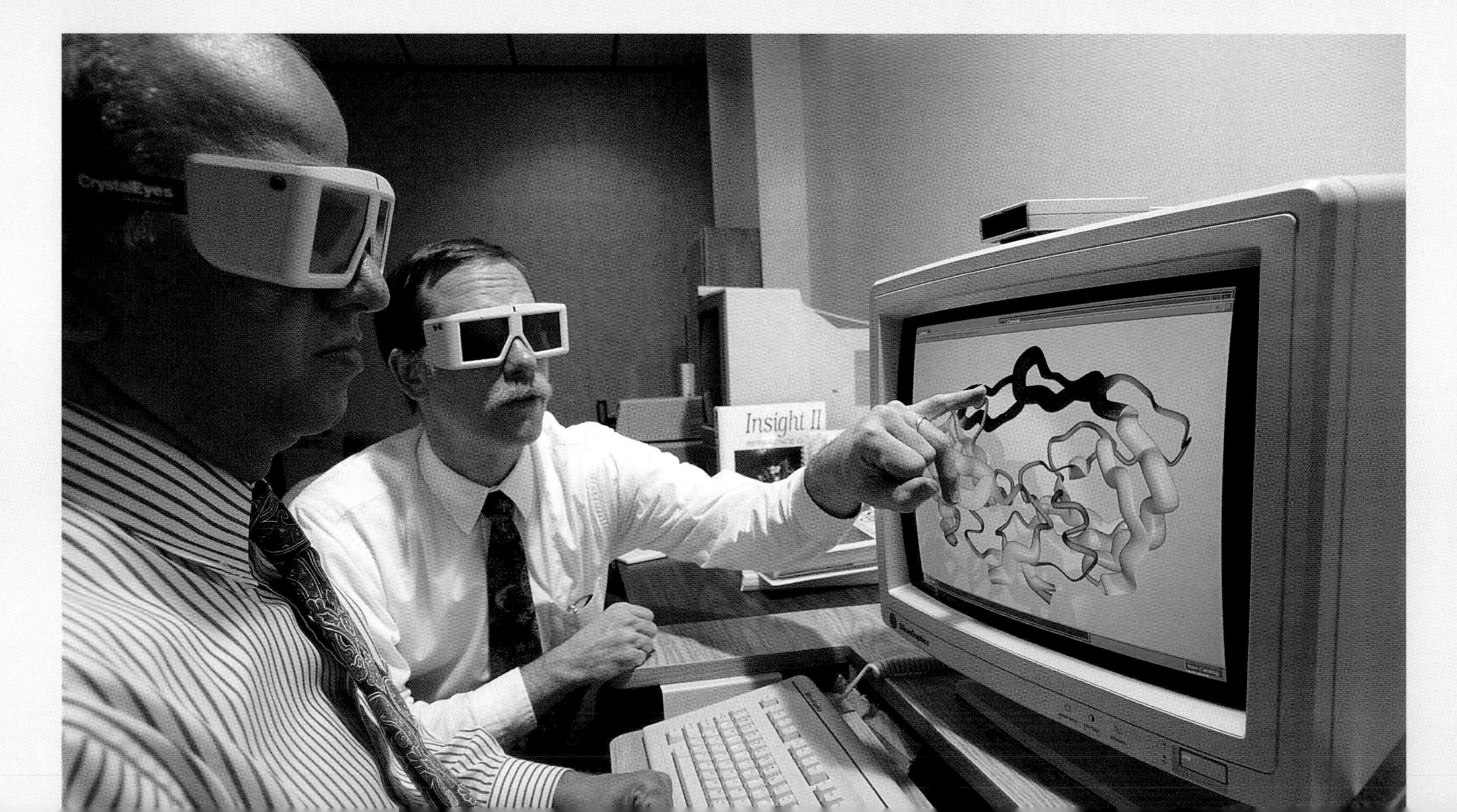


On the forefront of medical research, SAIC scientists investigate new pharmaceuticals and seek new antibiotics to fight drug-resistant and emerging new diseases.

moved from technical positions into management or marketing roles, SAIC's leaders recognized that the reputation and talents of the technical staff were key to winning new work. The ESTC provided a mechanism for individuals to take an increasing responsibility on a Companywide basis, without sacrificing their technical careers.

The ESTC provides a network that reaches out both within SAIC to ensure self-awareness of the company's technical capabilities, and outside of SAIC to the research community to ensure that the company remains on the forefront of emerging technology areas.

"SAIC's technical breadth and depth distinguishes us from other companies and certainly from our competitors for large information technology and systems integration programs," said Steve Rockwood, SAIC executive vice president with leadership responsibility for the Company's research and development work. "On many occasions we have been selected by customers because they knew that beyond performing the required work, we could help them stay on top of emerging technologies."

Since 1984, the ESTC has presented yearly awards for papers published in peer-reviewed journals. The selection criteria include technical quality, originality and uniqueness of approach and results, significance of results, contribution to the field, and effectiveness of presentation. Awards are given in the categories of Biochemistry/Molecular Biology, Applied Mathematics and Computer Sciences, Engineering and Systems Integration, Environment/Geophysics, Information and Communications Technology, Physics, Policy/Economics/Arms Control, and Technical Books.

"Offering awards for the best technical papers encourages our scientists to write about the important research they are conducting," said Chief Scientist Pam Surko, who co-chairs the ESTC with Bob Lucky, director of applied research at Telcordia Technologies. Surko is an expert in artificial intelligence, and has used neural networks for data mining for the CIA to predict genocide, ethnic war, revolutions, regime crises, and transitions from autocracy to democracy.

The ESTC's reach into the academic community includes sponsoring publication prizes for students at universities such as Georgia Tech, Howard University in Washington, D.C., San Diego State University, and the University of California, San Diego.

Many of the emerging commercial technology markets have their technical origins in research and development work conducted for the Department of Defense. By bringing the company's leading scientists together regularly, ESTC facilitates networking between its best scientists. This focus on both SAIC's technical breadth and its technical excellence helps maintain SAIC's prominence in the new technology landscape.

**PRINCIPLES & PRACTICES**

**SAI has always provided its employees with more freedom to pursue their interests and professional careers than do most other companies. All parts of SAI, even the more tightly managed entities, are far from the lockstep organizations with which we often find ourselves competing. Responsible individuals in the company use this freedom to conduct their business with a straightforward, generally decentralized approach. Avoidance of strict conformity, within SAI has fostered the growth of one of the strongest professional staffs ever assembled in the United States.**

***From "Principles & Practices of SAI," written by J. R. Beyster in 1983 to codify the Company's core values in a pamphlet for all employees.***

## High-Tech Warfare

In August 1990, Iraq invaded its neighbor Kuwait and prompted a regional conflict that led to an air and ground attack by the United States and its allies in February 1991. U.S. military officials spent a decade contemplating the prospects of a sophisticated global conflict, and this smaller Persian Gulf War benefited from the thorough preparation of soldiers and updated technology far more advanced than the guns and bombs of earlier wars. It was an unexpected, frontline test of the preparation of people and systems for a new style of modern warfare.

Most of the Army units had been through rigorous training at the National Training Center using systems designed and installed by SAIC in 1983 and updated in 1990. The simulations were realistic preparations for the high-tech warfare of Operation Desert Storm—the Tomahawk Cruise missile was highly accurate, and the Patriot Air Defense was successful in destroying Scud missiles launched by Iraq. SAIC assisted the Navy for many years in developing the Cruise missile and the Patriot systems; the training simulations contributed to battlefield performance. Surface ships and aircraft carriers were supported by SAIC's AMSEC subsidiary. AMSEC's technical staff and engineers flew from ship to ship by helicopter to assist crews in assuring the operational readiness of various ship systems. SAIC engineers and technicians remained in Saudi Arabia during the war to continue their support of the Saudi Navy's command and control system designed by SAIC well before the hostilities began. SAIC employees oversaw the successful operation of their system, and all survived the war.

After the end of the Cold War, SAIC's work with the Government continued to increase, though the nature of much of that work changed. Threat reduction and non-proliferation became international priorities, and regional conflicts were perceived as more imminent threats than global confrontations. SAIC was involved in numerous efforts supporting these revised national priorities.

In 1989, SAIC created the Center for Verification Research under a contract with the Defense Nuclear Agency and began participating in policy-analysis work that took on a new twist after the Cold War. Much of the policy analysis prior to 1990 focused on arms control, understanding the motives and intentions of superpowers, isolating satellite countries from the Soviet Union, and keeping distance between China and the U.S.S.R. After the Cold War, the work at the Center for Verification Research was directed toward non-proliferation of nuclear capability, threat reduction, and keeping nuclear, chemical, and biological capability away from groups or nations considered roguish in their desire to use those capabilities.

In support of international arms control agreements, the Center for Verification Research tests different technologies used in treaty verification.

SAIC provides safety, reliability, and quality-assurance support to two of NASA's most important efforts—operating the Space Shuttle and building the International Space Station.

While the verification and threat-reduction programs were being established, SAIC also devoted massive resources to building a simulation capability for defense preparation. The value of such training was clearly demonstrated in the Gulf War, and the Department of Defense chose to continue to enhance its simulation facilities.

SAIC constructed a distributive interactive simulation capability—distributive because people in different locations could participate simultaneously, and interactive because all participants got to see and react to a battlefield situation being portrayed. These participants could play their own real-life roles—a commander in Virginia, tank drivers in Kentucky, helicopter pilots in Louisiana, and fixed-wing aircraft pilots in Illinois—in ways sufficiently realistic to prepare all types of military participants for their roles in actual field operations.

For the U.S. Government, the simulation systems were a great money saver in a time of carefully monitored Government spending. Military experience could be honed without actually driving tanks or flying planes in every training effort, and military officials could determine the degree of compatibility of various mechanical and operating systems of the Army, Navy, and Air Force, as well as the compatibility of certain military systems used by United States' allies. For SAIC, these simulations represented a new application of its research and development and system's integration expertise.

During this time, SAIC also invested heavily in developing common approaches to system engineering and software development. By the late 1990s twelve SAIC operating units had achieved Level 3 of the capability maturity model of the Software Engineering Institute, a ranking only thirteen percent of assessed organizations achieve, and Telcordia had achieved Level 5 of software maturity which less than one percent of assessed organizations achieve. SAIC's strong engineering and software skills were the basis of major C4ISR (command, control, communications, computers, intelligence, surveillance, reconnaissance) systems integration support contracts at the Defense Information Systems Agency and the Navy Space and Naval Warfare System Command and software development contracts across military departments and federal agencies.

As national priorities shifted, SAIC adapted its basic skills to the new requirements, and during the decade from the mid-1980s to the mid-1990s, the Company's revenues from national security projects nearly tripled.

As part of the Company's Mobile Sea Range, portable shelters, left, allow operators to coordinate large-scale naval training exercises at sea. SAIC developed software to make the Close Combat Tactical Trainer, below, a realistic alternative to a field trainer.

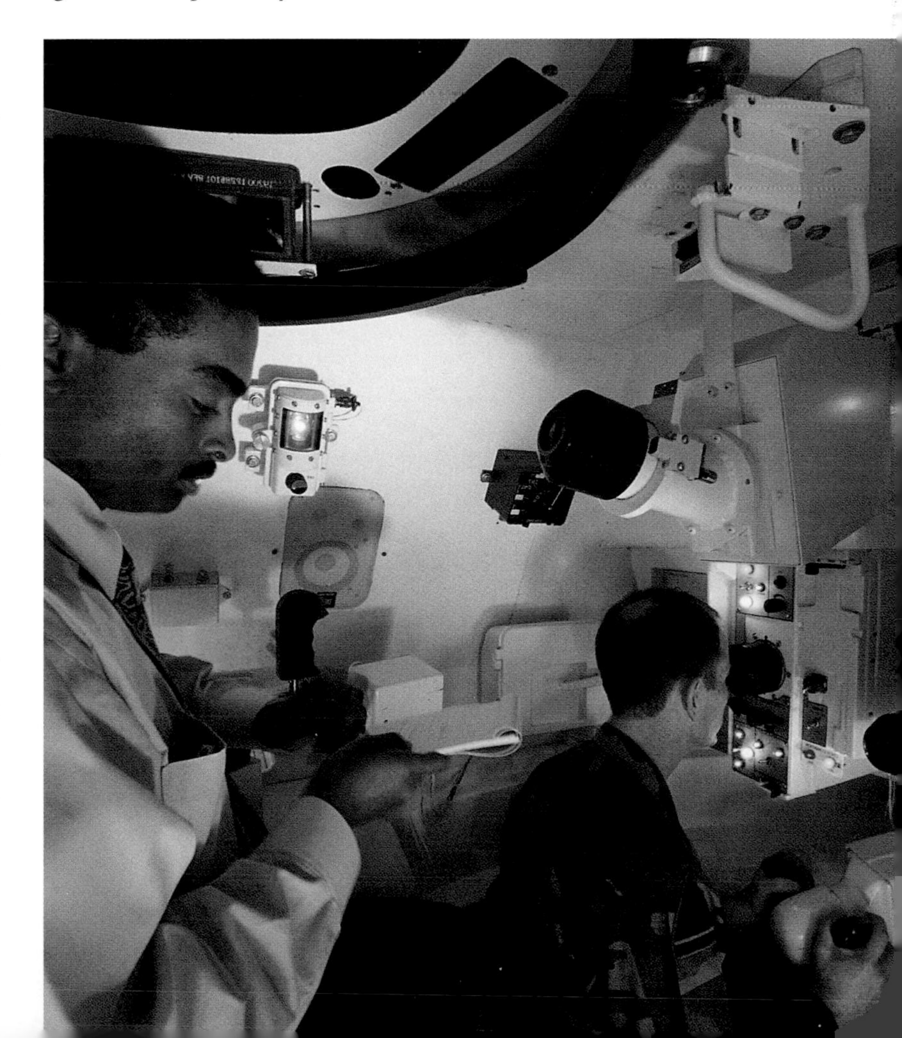

Inside a military pilot's helmet, SAIC's "Super Cockpit" technology displays a three dimensional representation of the real world, as well as essential flight data. For the pilot, this means less chance of error and increased odds for survival.

# Riding the Rails

The YardView system helps American President Lines at the Port of Los Angeles track, in real-time, hundreds of rail cars and thousands of intermodal containers, below and right.

For a motorist sitting at a railroad crossing and watching a freight train roll by, the experience can either be a moment of nostalgia or frustration, depending on the individual's perspective. But for the railroad companies, the perspective is focused on the critical need to keep track of all those railroad cars. Failing to know where every rail car is has huge financial implications for the rail carriers—lost freight or a rail car out of use costs big money.

The motorist at the crossing watching a locomotive pull a long string of cars probably does not realize that those cars don't all belong to the company that owns the locomotive—in fact, most of them belong to other public or private railroad companies, cars that are hitching a ride with another carrier. And that's the challenge.

Because it is common practice to mingle freight cars throughout the railroad system to get shipments delivered as quickly as possible from origin to destination, tracking the location of freight cars is a massive job. Owners want to know where their shipments are at any point in time. Billing each other for carrying these rail cars requires prompt and accurate information. A lost rail car is an asset out of commission, that costs the owner down time and tracking efforts.

In 1993, SAIC was part of an effort to develop a standard system for tracking railroad cars for the North American railroad industry.

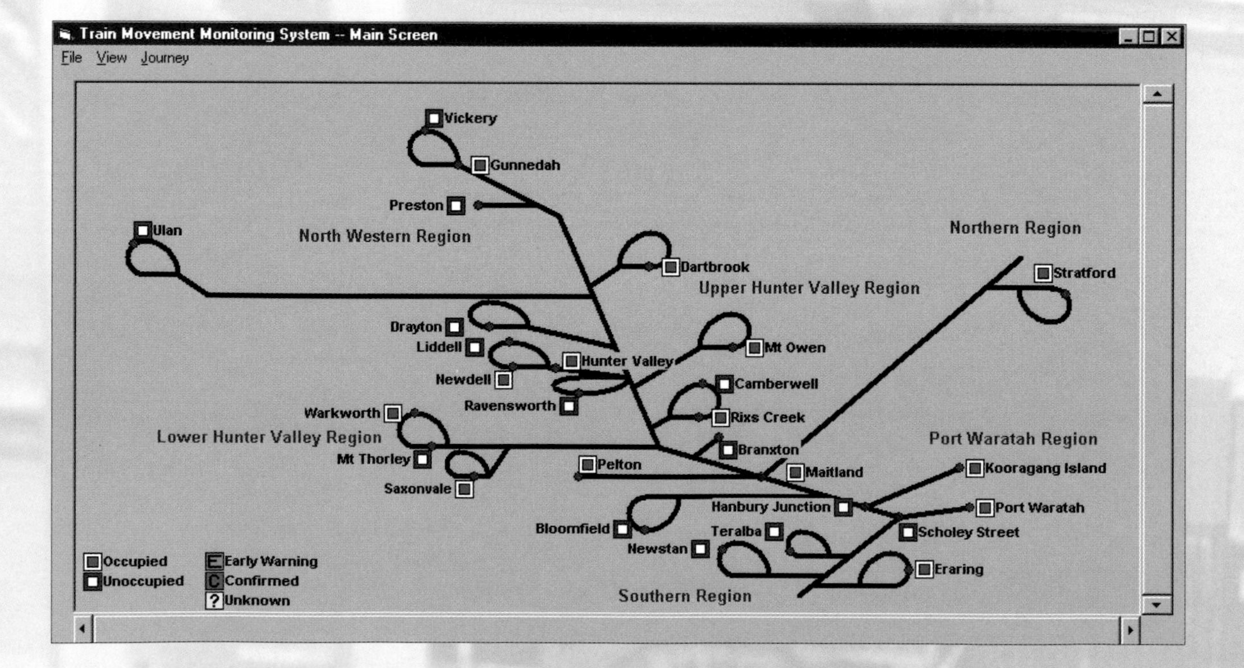

That system involved a technology previously used by SAIC in its toll-collection work on turnpikes and toll roads in the United States and Asia—a tag on the vehicle to identify its location and to transmit data for billing purposes. For the railroad industry, the system involved the installation of tags eight inches by two inches and five-eighths inch thick on each rail car. These radio frequency identification tags can be read by a scanner, similar to the tags retail department stores put on clothing to make certain inventory does not disappear through the door without being paid for.

For the railroad cars, the system involved an added complexity. Each car—and there are approximately one and a half million rail cars in North America—required its own identification code. SAIC worked with the company that manufactured the tags, and together they developed the system used to tag virtually every rail car. Approximately 5,000 scanner sites throughout North America read the tags as rail cars roll by, and these sites transmit data directly to the railroad companies, letting them know where the cars are on a continual basis.

This system was developed at SAIC by applying mature technology that had been successfully used elsewhere (on the toll roads) to solve an important problem for a new customer.

# Performance from Experience

The acquisition of Telcordia Technologies, Inc. (formerly Bellcore) in 1997 was a truly defining moment in the coming of age of SAIC. It was more than ten times larger than any other acquisition SAIC had made previously in terms of revenues, employees, and purchase price. In one stroke, SAIC became a material player in the commercial technology marketplace and specifically in telecommunications software and services—a huge and rapidly growing marketplace.

Telcordia is an SAIC company providing software, engineering, consulting and training services to optimize the performance of communications networks worldwide. Telcordia was created in 1984 as Bell Communications Research, or Bellcore, as a result of the breakup of AT&T to provide technological expertise and innovation for the RBOCs (regional Bell operating companies) in local exchange services, in contrast with AT&T, which would pursue strictly long distance services.

By the mid-1990s, the seven RBOCs, each owning an equal share of Telcordia, found themselves in competition with each other. They no longer desired to jointly own Telcordia, nor did they want it to be acquired by a company that might split it up.

Today, 80 percent of the U.S. telecommunications network depends on software that was invented, developed, implemented, or maintained by Telcordia personnel. The company's other critical function was to provide the U.S. Government with national security and emergency preparedness support. Telcordia innovations that benefited the RBOCs included Caller ID, Call Waiting, and the 800 toll free service. In addition to these customer services that have become cornerstones for marketing done by the RBOCs and other telecommunications providers, Telcordia provides 100 million lines of software to support the daily operations of telephone companies across the U.S. Telcordia has provided its customer's with network architecture, systems design, operations, reliability, and security—issues that are relevant for the twenty-first century.

As the RBOCs began looking for a suitable buyer for Telcordia, SAIC considered making a bid. At first, it was hailed a crazy idea. After all, SAIC had always enjoyed a strong cash position. This was a financially conservative company that had paid no more than $30 million for any acquisition prior to that point. Buying a company with $1 billion in revenues when SAIC was approaching $3 billion seemed ludicrous. David Overskei, a corporate strategic planner, was persistent in pursuing this opportunity. He would have to convince SAIC's leaders that Telcordia was a strategic fit. Bob Beyster, CEO, agreed to fly to New Jersey to meet some of the Telcordia employees. This nuclear physicist felt right at home. His

Research at Telcordia Technologies has shown that the capabilities of existing copper wires can be boosted by using ADSL—asymmetric digital subscriber line—with relatively little capital investment to meet the data-carrying demands caused by the “Internet frenzy.”

Telcordia Sapphyre(SM) loop qualification service helps telecommunications providers automatically determine whether asymmetric digital subscriber loop (ADSL) technology can help boost the data-carrying capabilities of existing telephone lines in a specific service area.

Sapphyre is a service mark of Telcordia Technologies, Inc.

mind began spinning with business opportunities the two companies could tackle together.

The underlying rationale for this combination was the convergence of information technology and telecommunications—both businesses and individuals increasingly rely on the availability of information in various locations on demand. In addition, each company had customers who were prospects for the other—SAIC's base of commercial and Government customers would likely need telecommunications work, and Telcordia customers were ideal targets for SAIC's research and development, systems integration and IT expertise.

SAIC made a bid to purchase Telcordia in 1996, and the regional Bells accepted, seeing SAIC as a fitting long-term partner. Together with Overskei and Beyster, CFO Bill Roper, then SAIC Treasurer (now Telcordia CFO) Ward Reed, Assistant General Counsel Kevin Werner, and Senior Vice President for Business Development Jim Idell worked tirelessly throughout the one year due diligence period to push through the regulatory approvals and consummate the deal. The team convinced the sellers that SAIC, with its technical breadth and depth, was best suited to shepherd the RBOCs critical software and systems assets. When SAIC employees heard that SAIC intended to purchase a company with $1 billion in revenues, concerns arose as to whether SAIC was jeopardizing its employee ownership by taking on significant debt. Employee owners had to be convinced that SAIC was on firm financial footing and that some level of debt is financially efficient.

The next challenge was introducing the 5,500 Telcordia employees to SAIC's unique form of employee ownership. Perhaps influenced by their close proximity to Wall Street, Telcordia's employees were financially savvy. When they first heard SAIC was the potential buyer, many raced to the SEC web site to review SAIC's financial filings (although SAIC is not publicly traded, it has more than 500 shareholders and thus must file with the SEC).

When the Telcordia employees first heard of SAIC's internal market, many were a bit skeptical. But as they came to know SAIC's leaders and managers over the one year due diligence period, skepticism turned to anticipation. Within one year of the sale closing, Telcordia employees invested approximately $165 million in SAIC, primarily through their retirement accounts.

As a result of its brand identity in the commercial marketplace and rankings in commercial industry publications, Telcordia raised SAIC's public profile. Through this partnership, SAIC would become more adept at winning commercial business. The addition of Telcordia in the SAIC family of companies also dramatically increased SAIC shareholder value.

# THE PATH TO THE FUTURE

*March 9, 1876: "The apparatus suggested yesterday was made and tried this afternoon . . . When Mr. Watson talked into the box an indistinct mumbling was heard. . . ."*

*March 10, 1876: "The improved instrument . . . was constructed this morning and tried this evening . . . I then shouted into [the mouthpiece] the following sentence: 'Mr. Watson–Come here–I want to see you.' To my delight he came and declared that he had heard and understood what I said."*

*Experiments made by A. Graham Bell, volume 1, pages 39 to 41*
*(The notebooks of Alexander Graham Bell)*

SAIC dramatically strengthened its position in both telecommunications and the commercial market with the acquisition of Telcordia Technologies (formerly Bellcore) in November 1997. Today Telcordia helps build advanced networks for a variety of customers. Most of those networks combine fiber optics with existing customer systems, right, to increase flexibility and contain costs.

March 9, 1999: "Bellcore, the telecommunications pioneer responsible for designing the modern U.S. telecommunications system, today begins operating as Telcordia Technologies—a new name reflecting the company's expertise in communications technology and its ability to reach 'accord' among major communications service providers that allows seamless network interconnection."

*News release announcing the name change of Bellcore to Telcordia Technologies 18 months after it was acquired by SAIC.*

SAIC acquired Telcordia Technologies (formerly Bellcore) in anticipation of a fundamental shift in the array of telecommunications products, technologies, and services that will, in many ways, be as startlingly new and different as the work of Alexander Graham Bell in the nineteenth century. These fundamentally new ways of communicating involve the convergence of voice, data, and video networks, as well as new ways of using the Internet and other emerging networks—all the result of persistent research by Telcordia.

In the 1980s and early 1990s, Bellcore developed many of the operating systems used by AT&T and today's regional Bell telephone companies. Moving into the new century, Telcordia is designing substantial segments of the complex global telecommunications infrastructure that is rapidly becoming the foundation for a new age of information and communication. Just as Alexander Graham Bell's technology upstaged the telegraph lines with a new form of voice communication, the systems evolving today will provide more choices for every user of communication systems. (Mr. Bell would perhaps be intrigued that the data in his notebooks is housed on the Internet and can now be transmitted to millions of users over communication lines originally designed for voice.)

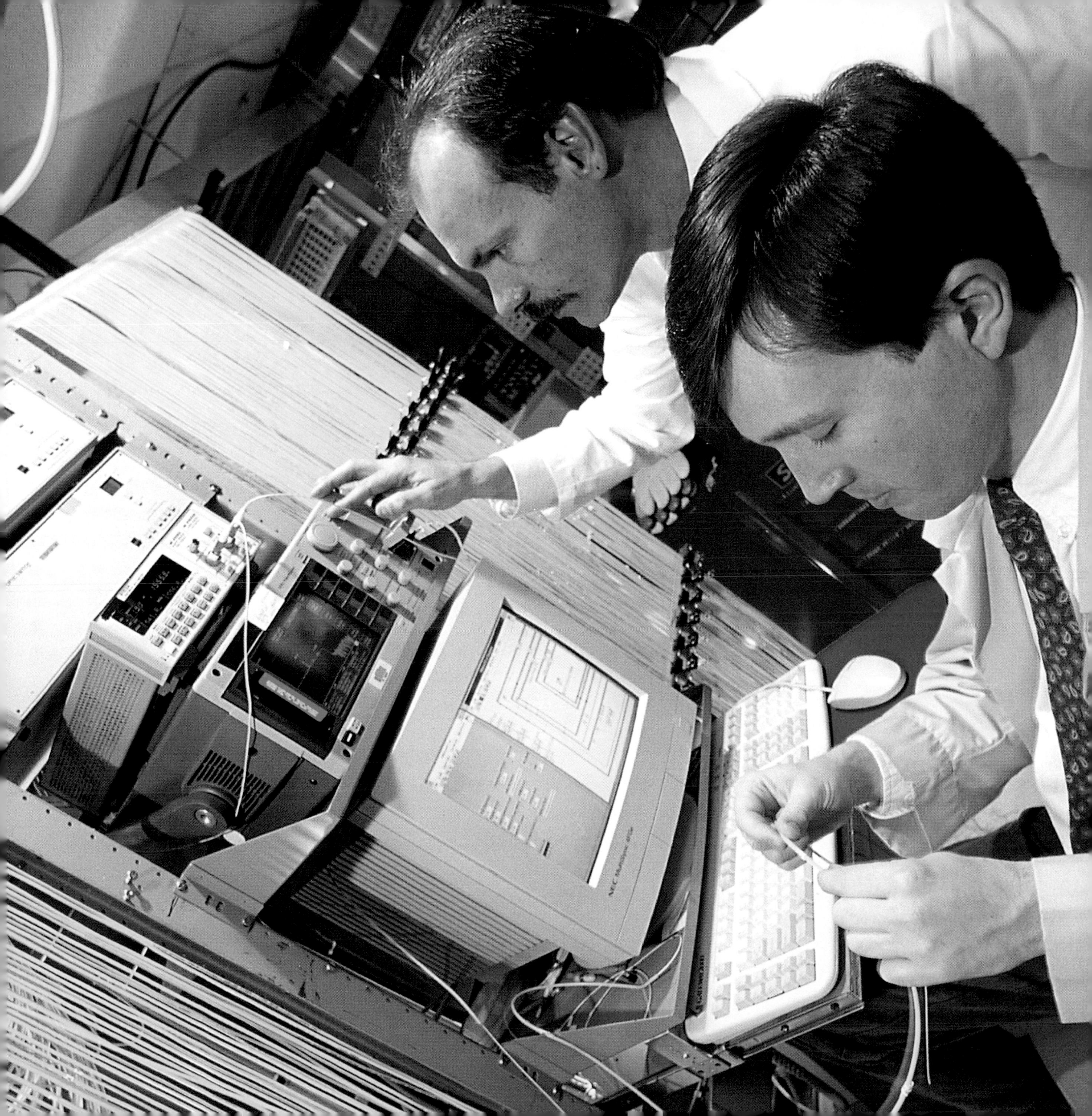

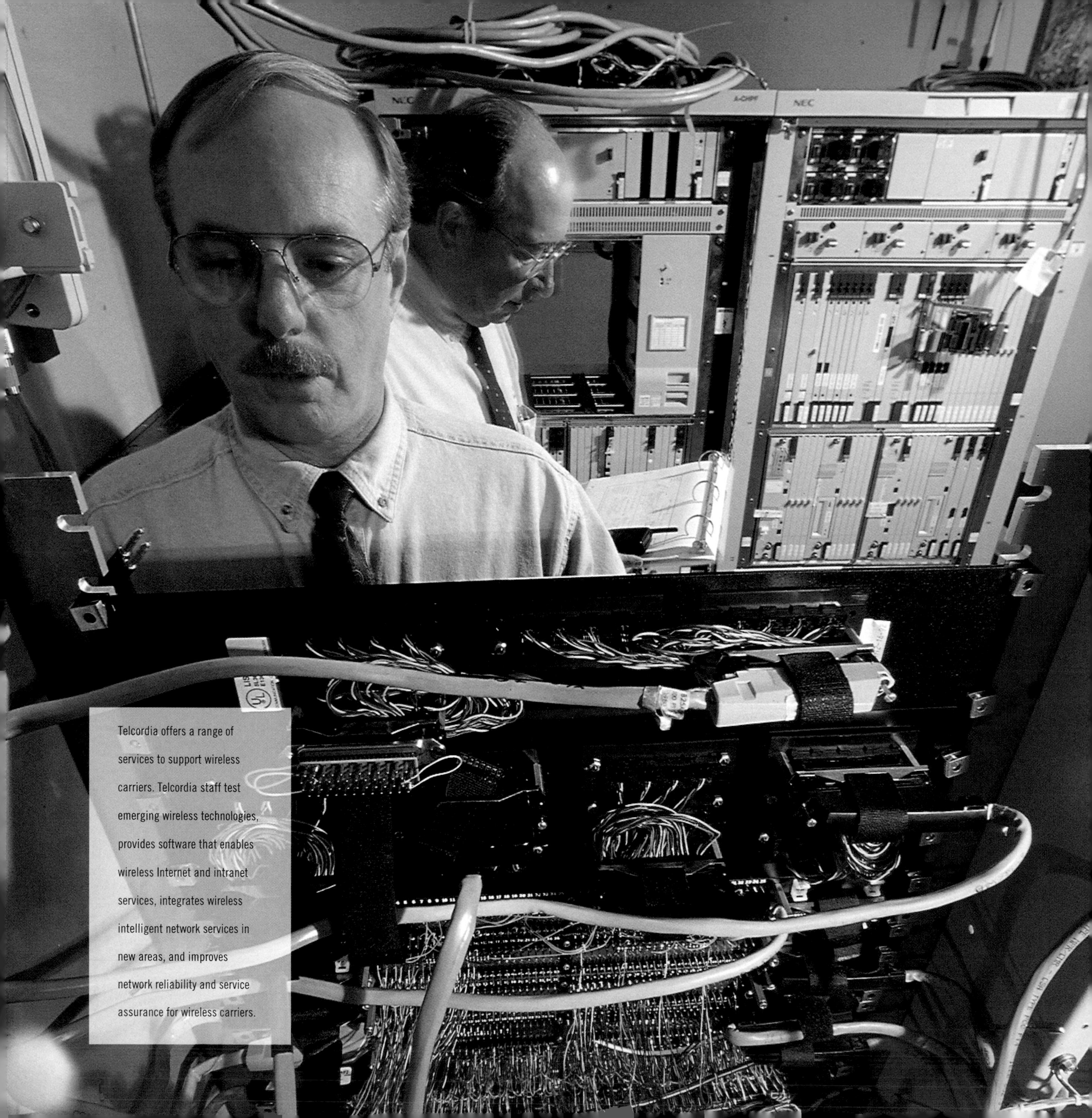

Telcordia offers a range of services to support wireless carriers. Telcordia staff test emerging wireless technologies, provides software that enables wireless Internet and intranet services, integrates wireless intelligent network services in new areas, and improves network reliability and service assurance for wireless carriers.

## Building a New System

The Telcordia customer base includes numerous local, long distance, and international telephone companies, as well as commercial and Governmental customers, and other telecommunications companies. For the telephone companies, Telcordia provides crucial assistance in raising performance and reducing costs as existing networks are modified to meet new requirements of customers and as technology changes.

Telcordia provides services that include the design and development of full-scale telecommunications systems, and with SAIC, it can address the integration of information technology and telecommunications. The Department of Defense and numerous civilian Governmental agencies used SAIC and Telcordia consulting services as they evaluated new information and telecommunications systems for various parts of the federal Government.

Telcordia assists telecommunications companies in modifying existing systems to accommodate important parts of new systems that can be adapted for their use. Some communications companies rely on Telcordia for the design and implementation of new networks, leapfrogging existing systems in an effort to quickly boost their competitiveness.

Wireless technology could be the answer to rapid and efficient deployment of such broadband services as Internet access and two-way multimedia teleconferencing to residential and business customers, left. Telcordia staff helped pioneer key telecommunications technologies such as asymmetric digital subscriber line (ADSL), right, which boosts the capacity of copper wire.

**Next Generation Networks: New Ways to Stay in Touch**

"This truly is the Big Bang that expands the universe of what telecommunications can do in our homes and businesses," said William Esrey, Chairman and Chief Executive Officer of Sprint, as he announced technological breakthroughs in 1998 that would provide Sprint's customers with high-speed, multi-function capabilities over a single telephone line, building on Telcordia technology.

Telcordia has world-class testing facilities to analyze new products in a realistic environment, right. The Telcordia training facilities, left, enable clients to learn how to use Telcordia-developed systems most efficiently.

These breakthroughs will allow a household or business to use one phone line to conduct multiple phone calls, receive faxes, and use the Internet at speeds up to 100 times faster than conventional modems. The need for multiple phone lines will be eliminated, video calls will be readily accessible and less costly, and use of the Internet will be so fast that typical pages on the World Wide Web will pop up almost instantaneously.

"We are opening new vistas for the ways in which people communicate. If you are a Sprint customer, you will be online, all the time. You will not have to access this network of breathtaking power and speed; you will be part of it," said Esrey.

Telcordia is providing the central software framework that is the core intelligence of Sprint's new system. At the heart of that software framework is Telcordia Next Generation Network (NGN). The NGN has grown out of years of research, close attention to the needs of established phone companies as well as newly emerging companies, and an awareness of the fundamental change the Internet has created in the world of communications and information technology. Historically, Telcordia has worked with its customers to add new services (such as Caller ID and call waiting) one at a time. With NGNs, telephone company customers of Telcordia will offer existing services in a less expensive way, will offer new services more quickly, and all services will be delivered in a fundamentally different way.

At the Telcordia ADSL Lab, researchers study ways to use this technology to dramatically increase the capacity of copper wire networks and enable high-speed Internet access. This technology is also used in lithium batteries for computers and other consumer products that require power and portability.

One of the most critical components of the NGN is the new Telcordia intelligent software program known as the Call Agent. This software acts like a switch in directing calls where they need to go. By moving this directing function out of the hardware and into the software, these software call agents give the telephone companies more control over the ways in which their networks are used, and they take far less space than the hardware of the circuit switches they replace.

The Call Agent software has two fundamental aspects. First, it enables phone companies to provide current services at much lower cost. Second, the flexibility of its design makes possible future services that stretch the imagination of both users and designers.

Cost reduction results from the shift of a core function from hardware to software, a change anticipated by Telcordia, building on its knowledge of the current U.S. telephone system, and its configuration, strengths, and weaknesses. In addition, SAIC's capability as a systems integrator helps drive down costs as telecommunications and information technology intersect.

The flexibility of an NGN results from its being a computing and communications platform that, for the customer, is everywhere at once. No longer limited by the old hardware switch that allows only one user or function by at a time on a line, it will allow numerous uses simultaneously. Its intelligent core will not only be smart enough to manage these complex gatekeeping functions; it will also become generally aware of the preferences and patterns of the customer. Managing voice mail and e-mail, facilitating e-commerce, communicating to turn appliances on or off, and providing the office work environment at home are all easy to imagine and configure with NGNs.

Underlying this technology is the question of how communications companies see themselves. Telcordia technology enables communications customers to become participants in the information-based lives of households and businesses, rather than commodity conveyors of voice and data. The Call Agent software and its flexible configuration allow the telecommunications company to determine precisely what their customers want and need, and to deliver it to them.

Telcordia and SAIC investments in Tellium, Inc., a subsidiary of Telcordia, are aimed at developing new optical networking technology designed to dramatically increase the capacity of fiber-optic next generation networks.

The advent of the NGN begins a journey to a new world for the telecommunications industry, and Telcordia and SAIC will be guides for their customers on the journey. SAIC has skills in developing large systems programs, and Telcordia has excelled with focused software development (such as Caller ID). The combined impact of the capabilities of these two companies bodes well for their journey together into the brave new world of telecommunications and information technology.

# The Marketplace of e-Business

SAIC has committed to employing its technical expertise to build user-friendly, highly secure, high-availability infrastructure to enable nonstop e-business and create virtual businesses into the twenty-first century.

By bringing together and leveraging its broad technical capabilities, SAIC entered the e-business marketplace to assist its clients in transforming their business models and operations through the use of science and technology. The deep security skills possessed by SAIC and Global Integrity, the sophisticated networking capabilities and scalable software solutions resident within SAIC and Telcordia, and the supply-chain and procurement applications, together with the hosting and Internet/Web directory expertise of SAIC and Network Solutions have been harnessed to deliver e-business solutions for SAIC's clients. The Company has created a virtual re-insurance exchange for the insurance industry, developed security plans for many large banks, and maintained Internet domain names for users worldwide.

In the telecommunications industry, SAIC and Telcordia pioneered Next Generation Network solutions for Sprint ION and Le Groupe Videotron Ltee. in Canada, developed electronic bonding applications to enable retail telephone companies to seamlessly transact business with many wholesale telephone companies, and oversaw the development of the Automotive Network Exchange (ANX), an Internet-based supplier network for the big three automobile manufacturers. For governments, SAIC has implemented innovative e-business solutions to electronically enable health care transactions for the Veterans Administration, procurement transactions for the state of Massachusetts, and a supply-chain infrastructure to expedite the production of aircraft, automobiles, and space vehicles for the Department of Defense.

SAIC's Electronic Commerce Rapid Application Development Laboratory, known as the EC RAD Lab, can create and operate e-business ventures from first concept to full operation.

SAIC developed the intensified visible-light camera prototype to acquire and track missile targets.

## Addressing the Changing Priorities of Government Customers

The government marketplace will continue to be an important component of SAIC's future. Following the end of the Cold War in the early 1990s, the national security components of the federal budget declined sharply. In addition, the budget caps negotiated between the Congress and the White House to achieve a balanced budget placed firm limits on the discretionary funds available to both the defense and civil sectors.

These budget pressures caused some companies to abandon the federal market and others—primarily the aerospace companies that had depended on large hardware procurements—to target the information technology segment of the Government market. In spite of the turmoil in this market, SAIC expanded its market share in the shrinking federal market and looked for new opportunities at the state and local level.

SAIC is leading the way in applications development using speech recognition and natural language understanding, whereby a computer can understand and respond to complex questions through plausible inferences or inductive reasoning. Applications for this technology include replacing "mouse clicks," voice dialing in a wireless environment, information retrieval, and voice portals for the internet.

In the decades ahead the continued pressure on all levels of Government will be to do more with less while improving support to the citizen. Already these pressures have resulted in new "paperless" contracting with the Government and radically new approaches for the procurement of goods and services such as objective-based procurements and share-in-savings contracts. SAIC is adapting to these new business models and is making investments in the mastery of new technologies necessary for continued growth in the Government market.

The changing Government market will provide SAIC new opportunities for the outsourcing and privatization of activities previously done in Government such as desktop support, network operations, training, facilities management, and a range of other functional activities. As already demonstrated with several state governments, SAIC will continue to help all levels of Government become more efficient by providing them the tools that enable electronic commerce and electronic business solutions. SAIC is developing and testing these new tools of electronic commerce in a laboratory dedicated to understanding the continued evolution of Internet-based business processes and electronic commerce.

For example, to ensure the security required for distributed service-to-the-citizen, SAIC is developing with it partners a public key infrastructure (PKI) that provides real-time on-line validation of digital certificates so that digital identities can be verified at the time of an e-commerce or e-business transaction. SAIC is developing advanced search tools to enable an individual to quickly find the information he or she is seeking from anywhere on the globe. SAIC is also developing powerful natural language speech recognition capabilities so that Government (and commercial) customers will be able to interact with their computers from any location without requiring access to a keyboard and mouse. And, SAIC and Telcordia are positioned to help all levels of Government move onto the Next Generation Networks.

In addition to maintaining its leadership in the dynamic fields of information technology and telecommunications, SAIC will continue to develop advanced technologies for the national security community. For example, for the Defense Advanced Research Projects Agency (DARPA), SAIC will continue to explore imagery understanding and automatic target recognition, photonics devices, hybrid power vehicles, and robotics and unmanned vehicles.

SAIC is the systems integrator for the army's DEMO III program to employ autonomous vehicles, in place of manned vehicles for reconnaissance and forward scouting missions, thus reducing casualties.

**PRINCIPLES & PRACTICES**

**The organizational structure that seems to fit SAI best is one in which people who are interested in the same things work together. This occurs through general business area compatibility, a common customer base, or other factors. These factors important to our people and their customers combine to produce group organizational entities that are clearly not textbook in nature but have proven over time to be viable from a financial, marketing, and a professional satisfaction standpoint. Therefore, SAI's group structure will be allowed to evolve into the next phase of the company.**

*From "Principles & Practices of SAI," written by J. R. Beyster in 1983 to codify the Company's core values in a pamphlet for all employees.*

## Diagnosing New Solutions for Health Care

In 1995, SAIC surprised the high-tech world when it was chosen by Kaiser Permanente over other larger, better-known companies to design a strategic health care information system. At that time, the worldwide medical system (which SAIC had successfully designed and implemented for the Department of Defense) was nearly in full use, and the Company wanted to use its knowledge, skill, and resources to expand its health care business and become a general vendor, not limited to the government sector. Although SAIC's work with the Government system had been highly regarded within the Government, the Company had not yet become a major player with large commercial health care providers when it won the Kaiser Permanente contract.

Kaiser needed a firm to work with them who had managed a large, complex health care program, and SAIC's qualifications were boosted substantially by its Government work. Though that qualification was important to Kaiser, much of the contract work involved understanding Kaiser's business processes rather than just providing technologically superior access to medical records. The SAIC team understood how different parts of a hospital interacted with each other and guided the Kaiser staff in shifting from a mainframe computer-centric system to a distributed computing infrastructure.

SAIC listened to the Kaiser clinicians describe what did and did not work well, and then focused on establishing agreement within Kaiser about specific needs. In this way, SAIC led Kaiser toward solutions based on a strong technical foundation that could adapt to various products. SAIC did not propose a product solution—selling goods off the shelf to meet various needs—but the Company worked with Kaiser to identify needs and potential solutions, allowing considerable flexibility regarding the use of particular products. This approach was consistent with Kaiser's business needs and with its inclination to build a consensus within the company before making decisions.

Winning the Kaiser Permanente contract gave SAIC a large boost to its credibility on the commercial side of the health care business. Saint Luke's Hospital in Kansas City hired SAIC to link outpatient and hospital facilities, and all health care providers in the Canadian province of Saskatchewan were given access to electronic medical records with an SAIC-designed system. The Company gained additional customers, and as the business grew, SAIC developed a strong reputation, and health care providers began approaching the Company with contract opportunities. SAIC was no longer the dark horse in every contract competition.

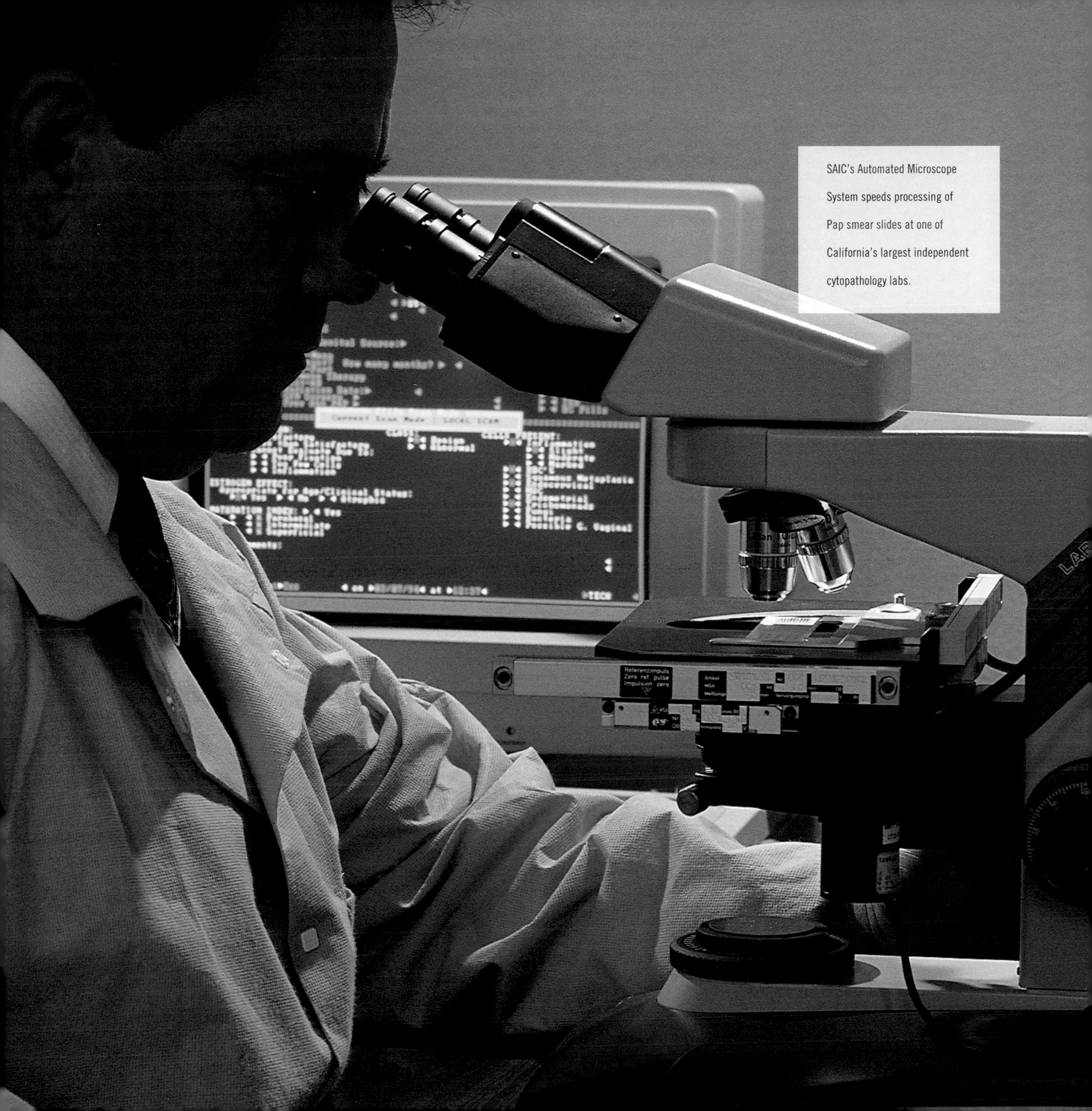


SAIC's Automated Microscope System speeds processing of Pap smear slides at one of California's largest independent cytopathology labs.

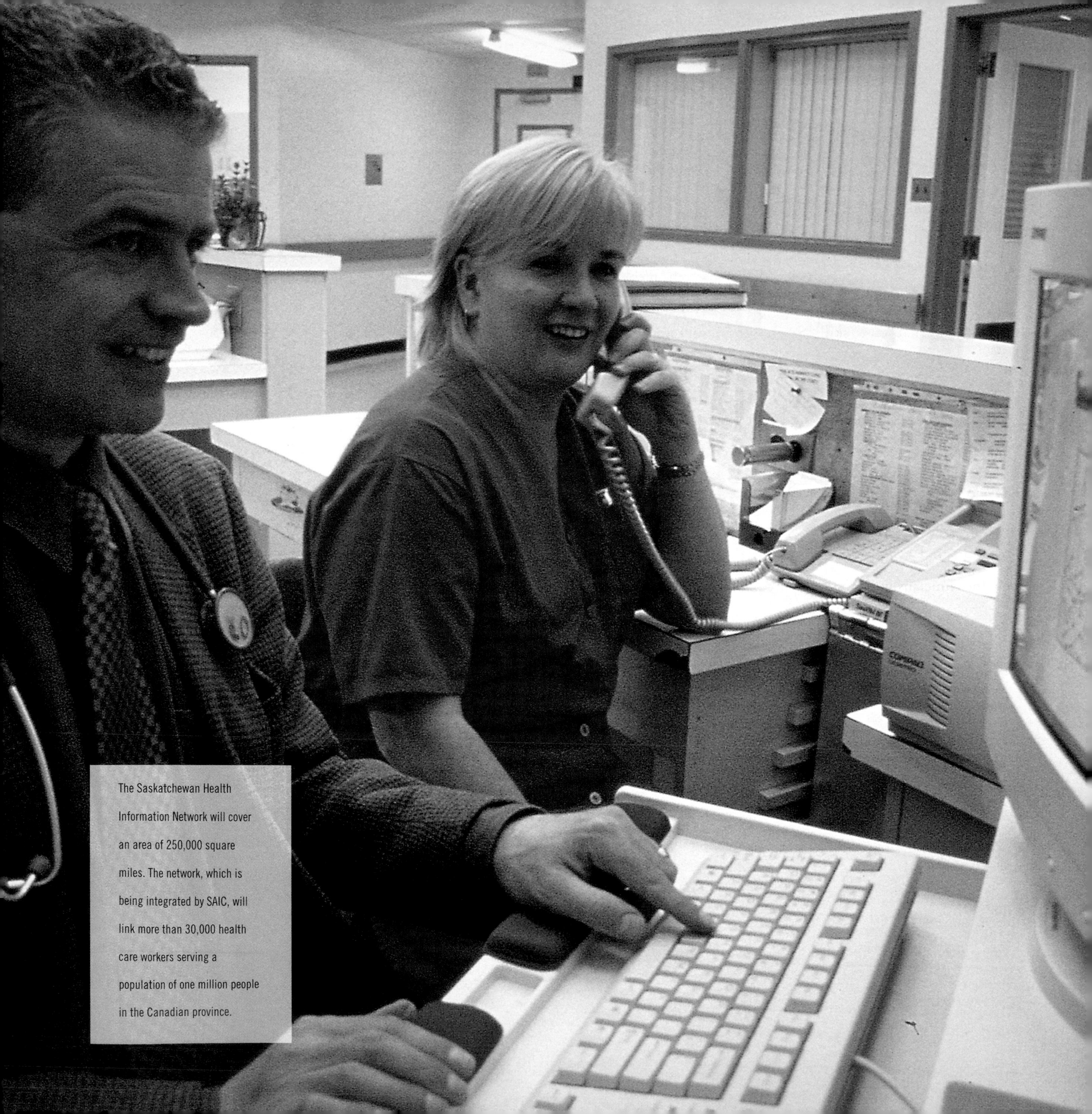

The Saskatchewan Health Information Network will cover an area of 250,000 square miles. The network, which is being integrated by SAIC, will link more than 30,000 health care workers serving a population of one million people in the Canadian province.

The fundamental approach SAIC has taken in its commercial health care business is that providers need systems integrators more than they need product vendors. SAIC's client-focused approach allowed the Company to design a flexible architecture to accommodate the unique aspects of each medical facility.

These systems integration skills have been the lifeblood of much of SAIC's business throughout its history, and they have been crucial in health care contracts. In building its commercial health care business on that core skill, SAIC taps resources that run deep, and the Company presents designs that allow consideration of all possible solutions and products. SAIC's open architecture approach allows customers to build on existing computer environments and incorporate new components as technology moves forward, balancing up-to-date functionality with cost concerns. In this way, SAIC differentiates itself from software providers who propose total solutions and hardware vendors whose various products do not readily fit together, and the Company provides greater value to its customers by focusing on how the clinicians want to run their company.

SAIC has begun incorporating work done by Telcordia in systems it designs for health care providers where geography is a concern. Kaiser Permanente operates in numerous locations that are distant from each other, and the health care providers in Saskatchewan are also distant from one another. Telcordia works with these customers on the architecture of telecommunications networks, and other parts of SAIC concentrate on network engineering and implementation of telecommunications systems.

Even for health care facilities not geographically distant, telecommunications networks can enable a physician to work in an office, at a hospital, or at home, and have the same access to medical records in any location. Internet-based communications can make the physician more accessible, less costly, and perhaps able to deliver higher-quality medical care through access to records and current standards of care.

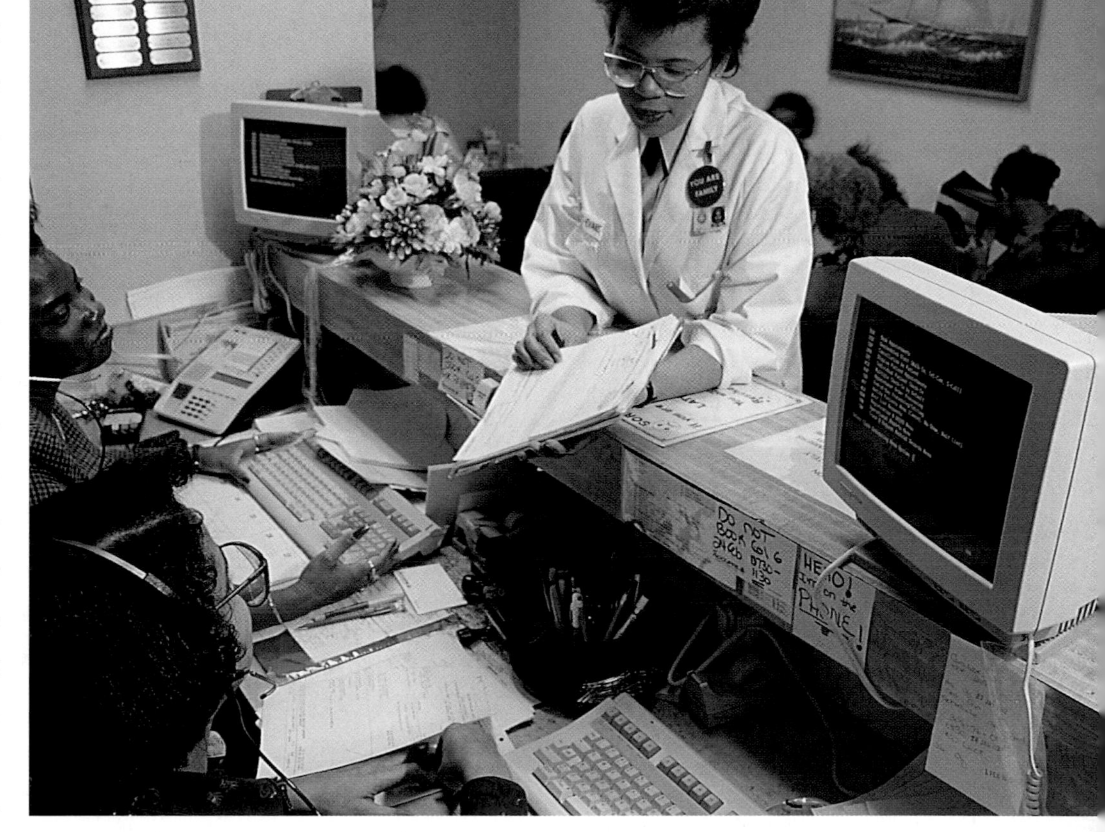

Administrators and doctors at Portsmouth Naval Hospital work in a new managed care environment using SAIC-developed systems.

# Information Protection

The benefits of universal, low-cost connectivity to customers, trading partners, and suppliers through the Internet are negated if critical information assets are compromised. Several SAIC organizations have world-class expertise in protecting critical information assets. SAIC's Global Integrity subsidiary helps 26 of the world's 100 largest enterprises protect their global networks and information systems.

Global Integrity helps companies and industries protect electronic payments and intellectual property. It also helped develop the Secure Electronic Transaction standard, proposed by Visa and Mastercard to secure payments over the Internet. SAIC's Center for Information Security Technology (CIST) implemented a Public Key Infastructure serving a six-million-user population to protect e-mail sent over the Internet and to enable secure transactions for various other applications.

Global Integrity, staffed by some of the world's leading information security specialists, maintains a long-standing commercial network incident response team, on call 24 hours a day. The team has prevented or recovered numerous losses.

Global Integrity is helping develop and operate an Information Sharing and Analysis Center, whose database enables financial institutions to work together to find and fix electronic security vulnerabilities. For a leading international financial organization, CIST deployed over 50 UNIX-based firewalls.

SAIC remains at the cutting edge of new information protection technologies through its own research and alliances. SAIC's alliance with network security leader ODS Networks, Inc. allows the joint offering of SAIC information protection expertise and ODS secure hardware and software, including the SAIC-developed Computer Misuse Detection System, one of the most sophisticated software tools for detecting security breaches by "insiders" (authorized users).

Telcordia is developing new techniques and technologies to detect attacks or failures on telephone and other telecommunications networks. The systems dynamically reconfigure networks to isolate compromised components, protect the hosts, and preserve network bandwidth.

SAIC's Center for Information Security Technology, below, monitors firewalls and security systems for clients around the world.

SAIC's Laser Line Scan made headlines when it was used to investigate the crashes of TWA Flight 800 in 1996 and Swiss Air 111 in 1998. Normally used for environmental surveys of the sea floor, the SAIC system can cover wider survey areas at very high resolution.

## Learning to Look Forward, But Not Straight Ahead

Over three decades, SAIC employees have built a strong, employee-owned company well-regarded by Government and commercial customers in a number of important domestic and international vertical markets. The largest vertical markets in terms of annual revenues in 1999 were national security, telecommunications, energy, health care, environment, and transportation. Other smaller but very important and growing vertical markets included financial services, space, criminal justice, and education. SAIC has grown these business areas and will continue to expand upon them by developing state-of-the-art technology through internally funded and customer-funded R&D activities; by facilitating knowledge-sharing across vertical markets and between Government and commercial activities; and by maintaining an environment that encourages entrepreneurial and innovative efforts by its employees.

While SAIC moved rapidly into commercial and international markets throughout the 1990s, the Company did not forget its origins. "We have a reputation second to none in Government circles," stated Chairman and Chief Executive Officer Bob Beyster in 1998. "We worked for 30 years to build it. Our future depends upon it. We view our Government expertise as symbiotic with our new commercial business. Each helps the other grow."

When the Company started doing defense-related projects, much of the expertise gained in that arena, such as nuclear reactions and nuclear safety, was ultimately used with commercial customers. Today much of the national security work benefits from projects done in the commercial sector. Information technology systems were initially developed to meet commercial needs, yet today Government users benefit from those developments in both defense and non-defense areas.

While SAIC's Government work continued to increase in absolute dollars in the late 1990s, the Company's commercial business grew at an even faster rate. In 1994, about 90 percent of SAIC's revenues came from Government business, and by 1999, nearly half the revenues came from commercial business. Part of this commercial growth was from internal sources, such as British Petroleum, health care, energy, and transportation. But another crucial part of the growth came from acquisitions and joint ventures—Network Solutions, INTESA, and Telcordia have provided substantial recent growth and building blocks for continued growth.

SAIC also seeks to maintain a balance between its traditional R&D based business (often involving smaller research contracts) and large undertakings (such as major information technology contracts).

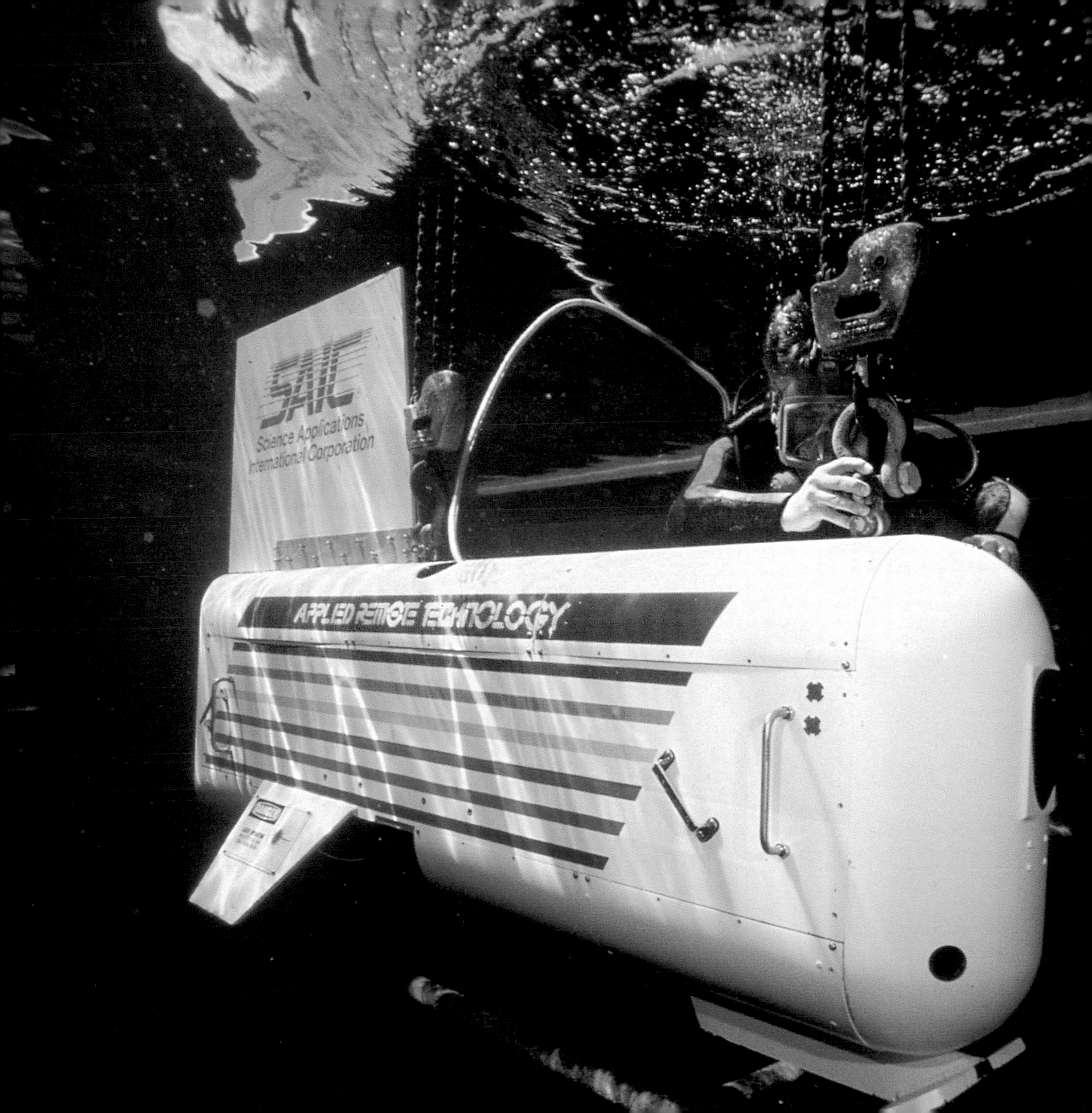
SAIC
Science Applications
International Corporation
APPLIED REMOTE TECHNOLOGY

SAIC's advanced techniques for assessing risk and devising cleanup goals have saved millions of dollars and won strong community support.

With markets changing at a breakneck pace, SAIC made major investments to position itself in the middle of emerging markets. "It seems like we're moving fast, but we have to be part of the technology revolution," said Beyster.

And yet looking beyond these strategic growth plans, underlying the relentless forward movement of the Company's business has been a particular passion of its employee-owners. They are a group of people who work together and share a common bond—ownership. As a group, they acknowledge that it is acceptable to argue among themselves about the pros and cons of a particular idea, but they recognize that they cannot disagree about working together.

For one of the largest contracts with the Environmental Protection Agency (EPA) in the Company's history, SAIC information systems helped the EPA develop solutions for entire ecological systems.

SAIC develops new, cost-effective technologies to improve water quality. The Safe Drinking Water Information System helps states monitor water quality at facilities. Below, at the Periphyton Water Garden in Orlando, Florida, an SAIC-developed algae-based filtration system cleans lake water.

Though the system works well, it is not perfect. Part of SAIC's formula has been a certain amount of organized chaos, perhaps in part from the scientific training of most of its leaders. Every year the Company's organizational chart is changed, creating different allocations of responsibility for different people, and aligning certain businesses to create more synergy and growth to position the Company to respond to anticipated changes in the world. Who knows what will happen next year, where the best opportunities will occur, and which customers will provide the best prospects for new business? If managers and employee-owners learn how to rapidly adapt to change, and are supported in working their way through it, they will be better able to navigate their way through the unexpected chaos of the world outside their offices. Rapid adaptation to change has become part of the Company's culture. It facilitates innovation—thinking "outside the box"—and it keeps people on their toes. "We work better when we're scared," is how Bob Beyster describes it. That attitude has kept SAIC healthy

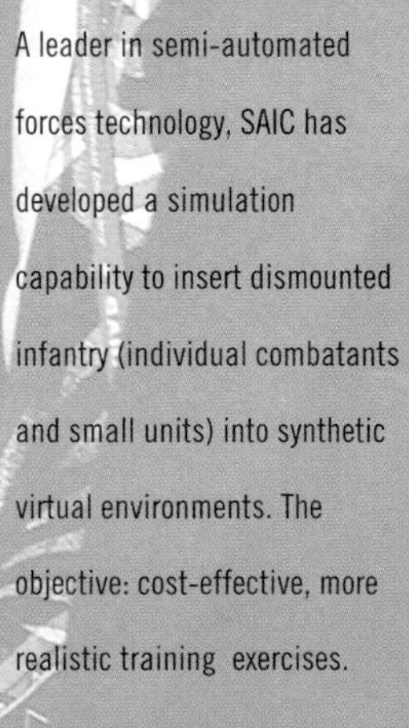

A leader in semi-automated forces technology, SAIC has developed a simulation capability to insert dismounted infantry (individual combatants and small units) into synthetic virtual environments. The objective: cost-effective, more realistic training exercises.

SAIC's information technology solutions are powerful weapons in today's law enforcement arsenal. An SAIC system gives federal, state, and local law enforcement agencies fast, on-line access to criminal histories, right. To support the "Brady Bill," call centers managed by SAIC run criminal background checks on prospective handgun buyers, below.

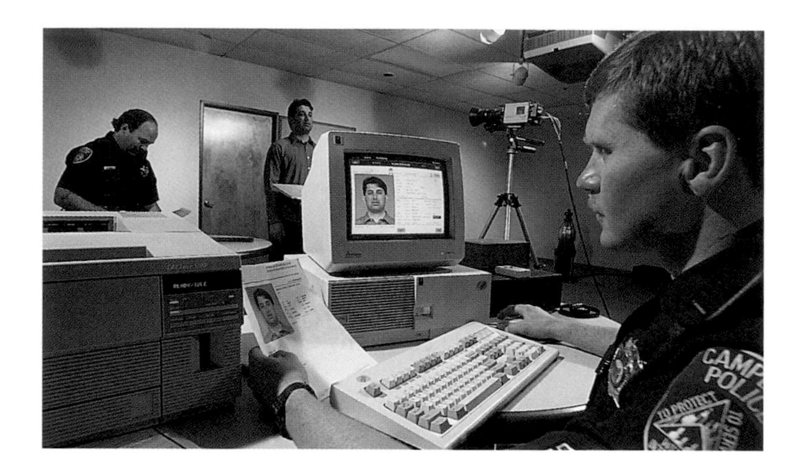

and growing through business, political, and technological changes that have disabled or destroyed competitors.

The Company's greatest strength is that while there is a common bond that holds these people together, there is no evidence that everyone thinks alike. History shows that controlled chaos inside the walls of SAIC has gotten the Company to where it is today: well placed on a sturdy foundation and ready to move forward on a number of different fronts. SAIC stills sees itself as a science and technology company, one full of internal synergies that feed each other and generate new ideas. In that way, Science Applications International Corporation is what it always has been and what it might become.

The latest technologies to help Latin American companies prepare for the new millennium are available at the INTESA Millennium Center, below. The center provides Year 2000 consulting, assessments, remediation and validation services, and technical support.

# Entrepreneurial Employee Ownership: Fueling SAIC's Growth

In 1999, SAIC celebrated 30 years of employee ownership. It is the third largest majority employee-owned company in the U.S. and the largest high-technology employee-owned firm in the country, perhaps the world. SAIC is recognized in business books and magazines for its unique approach to employee ownership. SAIC frequently hosts visits from leaders in business and government who hoped to gain an understanding of how SAIC creates both the reward structure and the entrepreneurial culture of ownership.

SAIC's employee-ownership system is the most comprehensive and complex in existence—the Company uses most every available mechanism to put ownership in the hands of those who are creating its success. No other company has an internal broker-dealer as sophisticated as Bull, Inc., which facilitates quarterly trades among buyers and sellers.

In 1999, some 95 percent of SAIC's 38,000 employees were owners. SAIC's top 45 elected officers and directors and their families held only 5 percent of the Company. This included Bob Beyster, who owned 1.4 percent of the Company.

Employees gain ownership through retirement plans, stock purchases, and stock incentives (a host of programs using stock options and stock bonuses as reward for performance). Changes to some aspect of SAIC's ownership system are made frequently in response to the changing capital needs of the Company, its work force, or the regulatory environment. This ever-evolving system requires a significant investment of time and money to develop and maintain over the years.

"The important thing to recognize about employee ownership is that it needs to change," said Beyster in a 1998 interview, "that the world is changing and the company is changing and the employee base of the company is changing. To sit statically by and try to always use the same kind of employee-ownership program is not going to bring success."

SAIC stock provided healthy returns to employee shareholders throughout the first 30 years. The stock price decreased only four times in three decades, and on every occasion it more than recovered the next quarter.

The latter 1990s brought challenges to employee ownership at SAIC. The Company had to become open to new forms of employee ownership. The INTESA joint venture in Venezuela tested employee ownership of SAIC through a new relationship with a customer outside of the U.S. The INTESA employees responded to employee ownership overwhelmingly positive—more than 80 percent purchased stock on

# SAIC Family of Companies

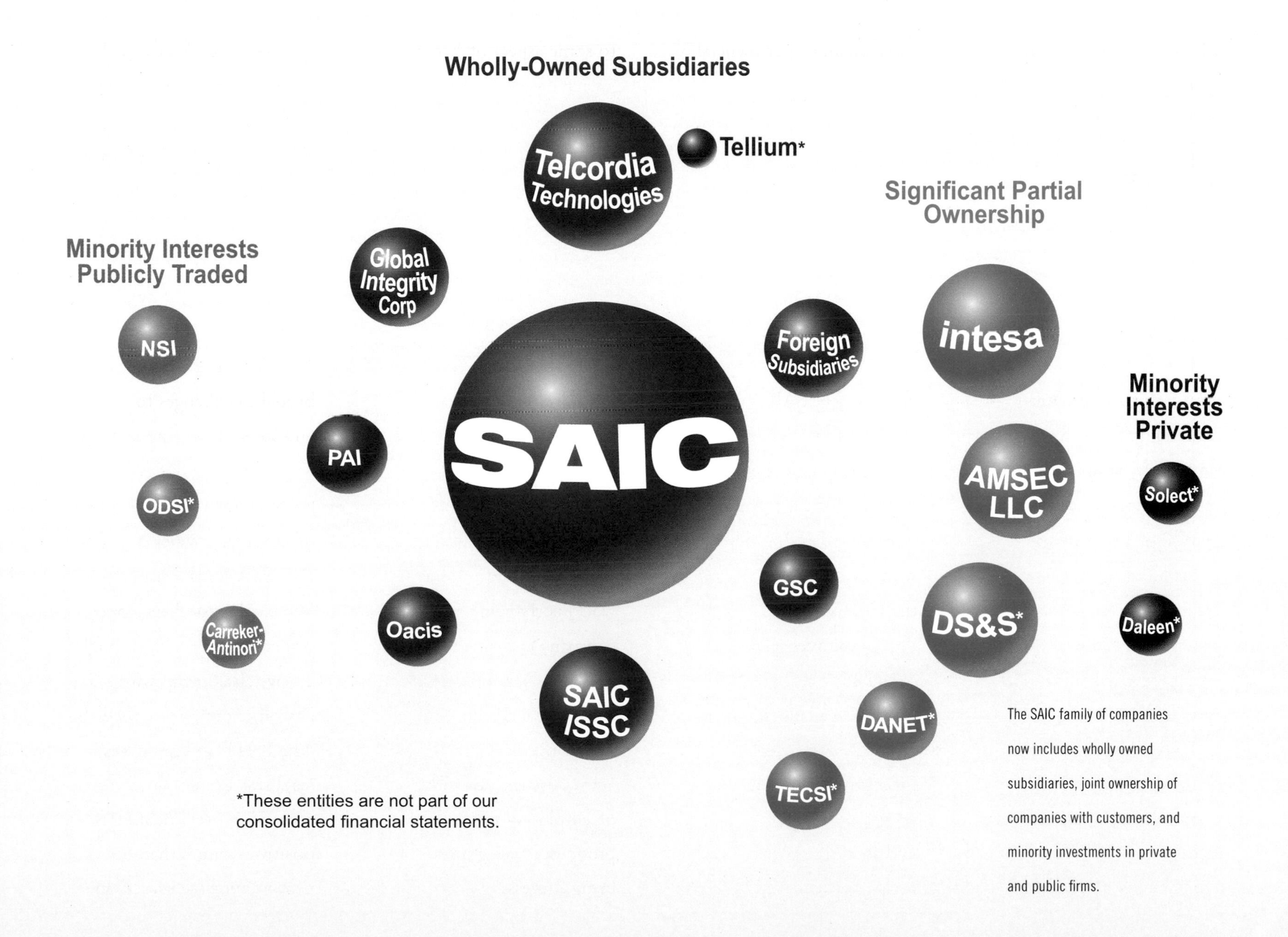


*These entities are not part of our consolidated financial statements.

The SAIC family of companies now includes wholly owned subsidiaries, joint ownership of companies with customers, and minority investments in private and public firms.

# SAIC Stock Price

## 1969–1999

SAIC's stock price has grown consistently over 30 years, achieving an annualized growth rate of 40 percent between 1994–1999 and an annualized growth rate of 25 percent between 1989–1999. This graphic does not reflect a 4-for-1 stock split that took effect August 31, 1999.

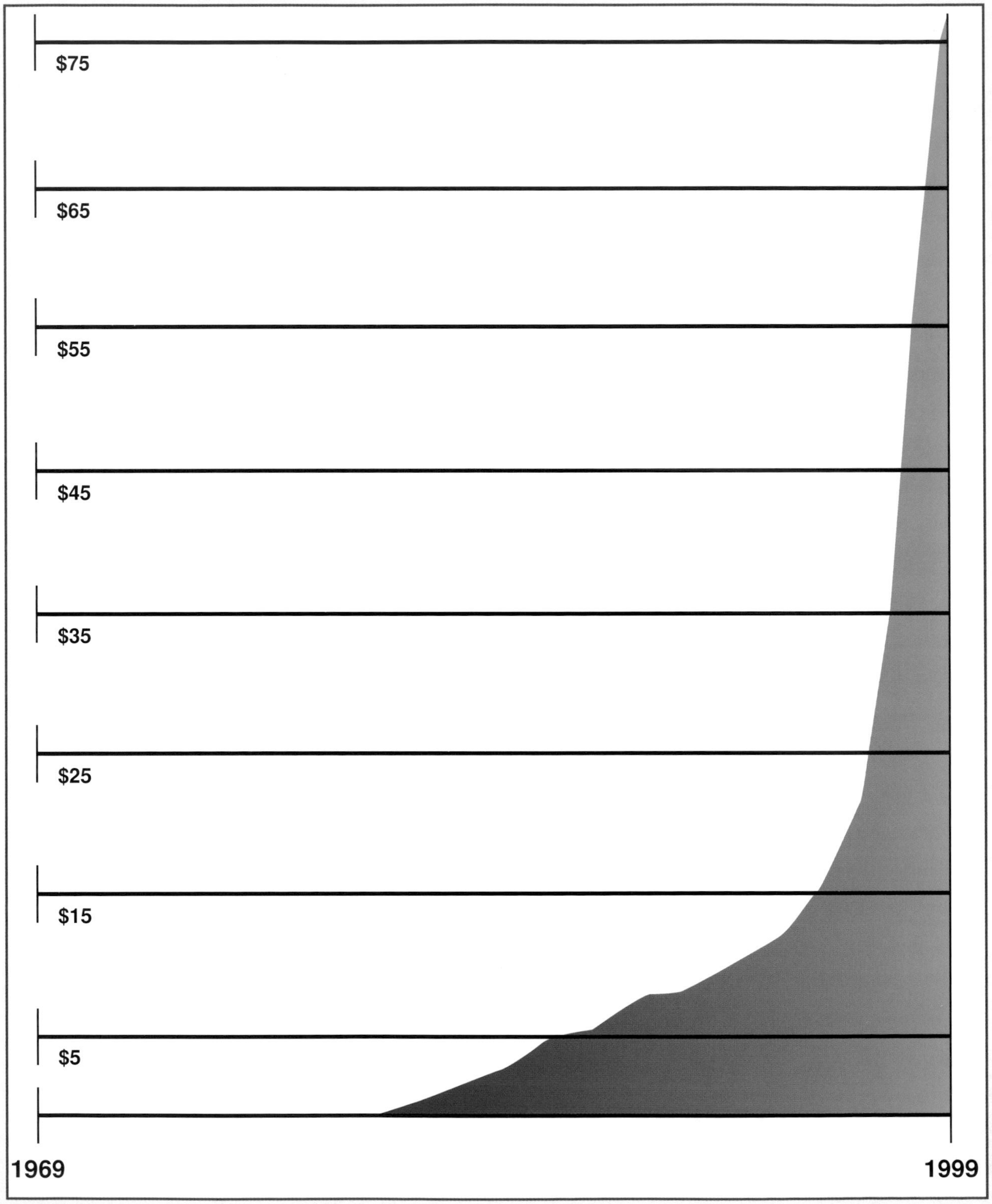

their first opportunity to do so. The INTESA employees adopted an ownership culture to leverage the psychological aspects of ownership.

The Telcordia acquisition in 1997 caused concern among SAIC employees since it was the first time SAIC anticipated taking on significant debt. They were also concerned about how the 5,500 Telcordia employees would respond to SAIC's unique brand of employee ownership. The Telcordia employees were cautious, asked lots of questions, and in the end, invested some $165 million in SAIC, predominately through their retirement plan balances.

Selling part of SAIC's ownership of Network Solutions, Inc. to the public in an Initial Public Offering also caused concerns among SAIC's employees. It was seen as contrary to the Company's founding philosophy and commitment over the years to remain private to ensure that the employees controlled the Company's destiny. The Company has addressed this by taking steps to ensure that NSI would establish an employee ownership plan of its own to share the equity fairly with employees.

Recognizing the need to ensure that all of SAIC's employees (over half had been with the Company for less than two years due to rapid growth) understand the benefits of being employee-owned, SAIC formed the Employee Ownership Working Group (EOWG) in 1997. The EOWG members (SAIC employees) are committed to ensuring that the employee ownership spirit remains alive and well at SAIC.

By 1999, SAIC had developed from one core company into a family of wholly owned subsidiaries, partially owned private companies, and partially owned public companies staffed by people sharing a common goal as owners of SAIC. Although SAIC's ownership system changed dramatically over 30 years, the Company's dedication to its founding philosophy remained tried and true: "Those who contribute to the Company should own it, and ownership should be commensurate with an individual's performance and contributions as much as feasible."

# Acknowledgments

When I first met Bob Beyster in his office, I told him that the poster behind his desk summed up SAIC, "None of us is as smart as all of us." His face shifted from intense concentration to a broad smile. "You've got that right," he said, and I knew I was beginning to understand how his vision had created SAIC and motivated thousands of people.

It has been a fascinating experience to understand the business and culture of this company, and to tell its story. The breadth of SAIC business is extraordinary—the continual efforts to bootstrap one technology into numerous applications is a commendable and highly successful strategy. Underlying the success of the business is the Company's unique culture of employee-ownership, a powerful part of Bob Beyster's vision that empowers employees to view the Company as their own.

I am grateful for the thoughtful help of Peggy Walkush in shepherding the production of this book from start to finish. She was very much involved in defining the structure and flow of the story early on, and as the writing progressed, she became essential to the entire process.

Numerous other SAIC managers have provided personal insights about various parts of the history and the business. Bill Layson and Matt Tobriner steered me through the early days and provided an understanding of how history laid the foundation for current and future business. John Glancy, John Warner, and Joe Walkush talked with me about today's business and the directions in which it is moving. Sam Carroll and Jim Idell filled many gaps in my understanding of particular projects and marketing activities, past and current.

Ed Straker walked me through the history of SAIC work in energy and environmental matters. Don Telage told me the story of Network Solutions, and Craig Cummings explained the various parts of the transportation business. Ken Sunday described the complex story of how SAIC became involved with the National Cancer Institute's Cancer Research and Development Center in Frederick, Maryland, and Paul Sager described SAIC's INTESA joint venture in Venezuela.

Beth Anderson, Ron Sullivan, and Dave Brooks gave me the past, present, and future picture of health care business at SAIC. Steve Rizzi explained the story of the Division Manager's Forum, and Bill Scott described the Technical Environment Committee.

Nola Smith was immensely helpful in explaining complex scientific matters in a way that I (not a trained scientist) could understand. Nola and Peggy Walkush both provided numerous written materials and video presentations to help with the writing of this story. I have drawn heavily upon those materials in this text and appreciate being able to use them.

I am grateful for the guidance of Tim Connolly at Tehabi Books, particularly in his coordination of the layout of the book with the written manuscript. This book was written and produced at a rapid pace, and I appreciate his continuing effort as well as the work of Laurie Gibson, who read each draft and helped refine and polish the work as fast as it came out of my printer.

—Stan Burns

# SAIC in 1999

### *Facts and Statistics*

- $4.7 billion in revenues
- $150 million in profits
- 38,000 employees
- 30 years of continuous growth in revenues, profits, and employees
- Market value of more than $4.5 billion
- 45% 5-year annualized stock price growth
- 25% 10-year annualized stock price growth
- 95% of employees are owners
- 80% of SAIC owned by current employees
- 5% of the Company owned by the top 45 elected officers and directors
- 79% of SAIC's professional staff hold degrees
- 47% of degree holders have advanced degrees
- 68% of degrees are in science or engineering
- 79% of degree holders have more than 10 years experience
- Offices in 150 cities worldwide
- Approximately 400 divisions worldwide with a great degree of autonomy
- 50% of revenues from government business; 50% from commercial business
- 6,000 active contracts
- 65% contract win rate (dollars won as a percent of dollars bid)

### *Rankings*

#1 Best R&D Company among Top 100 Private Companies
*Red Herring Magazine, 6/99*

#1 Private Info-Tech Company
*BusinessWeek, 6/99*

#2 Environmental Protection Agency Contractor
*Government Executive, 8/98*

#2 Health & Human Services Contractor
*Government Executive, 8/98*

#6 Top 20 Worldwide Development/Integration Companies
*Gartner Group, 11/98*

#8 Top 10 Government Prime Contractors in FY98
*Federal Computer Week, 5/99*

#8 Top 20 Worldwide Professional Services Companies
*Gartner Group, 11/98*

#9 Top 100 Defense Contractors
*Government Executive, 8/98*

#11 Top 50 Worldwide Service Companies
*Global Technology Business, 11/98*

#14 Top 20 Worldwide IT Consulting Services Companies
*Gartner Group, 11/98*

#347 Fortune 500
*Fortune, 4/99*

#34 Biggest Increase in Revenue of Fortune 500
*Fortune, 4/99*

#34 Biggest Increase in Profits of Fortune 500
*Fortune, 4/99*

An SAI scientist views the effects on titanium of a high-power carbon dioxide laser, one of several lasers that SAI will use to study laser-hardened materials for the U.S. Air Force.

# Strategic Locations 1999

***United States***

Albuquerque, NM
Anchorage, AK
Boston, MA
Charlotte, NC
Chicago, IL
Cincinnati, OH
Denver, CO
Durham, NC
Frederick, MD
Hampton Roads, VA
Harrisburg, PA
Honolulu, HI
Houston, TX
Huntsville, AL
Idaho Falls, ID
Indianapolis, IN
Kansas City, MO
Lisle, IL
Las Vegas, NV
Los Angeles, CA
McLean, VA
Morristown, NJ
Oak Ridge, TN
Oklahoma City, OK
Omaha, NE
Orlando, FL
Piscataway, NJ
Raleigh, NC
Red Bank, NJ
Sacramento, CA
San Antonio, TX
San Diego, CA
San Francisco, CA
Santa Barbara, CA
Seattle, WA
Tucson, AZ
Washington, D.C.

***International***

Argentina
Australia
Canada
Colombia
Czech Republic
Egypt
France
Germany
Ireland
Italy
Japan
Korea
Malaysia
Mexico
Phillippines
Saudi Arabia
Singapore
South Africa
Switzerland
Taiwan
United Kingdom
Venezuela

RING LASER
COHERENT
COHERENT

# Index

*Except as noted below, the photographs used in this book appear through the courtesy of SAIC.*

*Note: Letters refer to the position of the photo on the page, beginning from left to right and proceeding from top to bottom.*

| *Page(s)* | *Photo Credits* |
| --- | --- |
| 8, 120 | ©1994 by Bellcore |
| 14–15 | © Creation Creations, Inc. |
| 17 | ©1998 by Eric Millette |
| 18A | ©1998 by Eric Millette. Reprinted by permission of *Forbes* Magazine ©1999 *Forbes* 1997 |
| 18B | Courtesy Bob Beyster |
| 20A, 20B, 21, 23B | Courtesy of Bob and Betty Beyster |
| 26A | Courtesy of Bill Wright |
| 26B, 32, 33 | Courtesy of Betty J. Roberts |
| 27 | Courtesy of Cathy Fairchild |
| 26–27C, 36A | Courtesy of Constance Putnam |
| 28A | Courtesy of John W. Crabbe |
| 31 | Courtesy of William M. Layson |
| 36B | Courtesy of Dr. Robert E. Turner |
| 37 | Courtesy of Alan Holman |
| 42A | © Gary Mansur |
| 43A | Reprinted through the courtesy of the editors of *Fortune* ©1992, Time, Inc. |
| 44–45 | Courtesy of Wilson S. North |
| 102–103 | © Superstock |
| 118, 120–121 | ©1993 by Bellcore |
| 119 | ©1995 by Bellcore |
| 122–123 | *Gobal Networking* © 1998 by Jeff Brice |
| 126–127 | ©1994 by Bellcore |
| 141 | Courtesy of Sherry L. Taylor |
| 147 | ©1993 by Lee Peterson |